ASSESSMENT OF INERTIAL CONFINEMENT FUSION TARGETS

Panel on the Assessment of Inertial Confinement Fusion Targets

Board on Physics and Astronomy

Board on Energy and Environmental Systems

Division on Engineering and Physical Sciences

NATIONAL RESEARCH COUNCIL
OF THE NATIONAL ACADEMIES

THE NATIONAL ACADEMIES PRESS
Washington, D.C.
www.nap.edu

THE NATIONAL ACADEMIES PRESS 500 Fifth Street, NW Washington, DC 20001

NOTICE: The project that is the subject of this report was approved by the Governing Board of the National Research Council, whose members are drawn from the councils of the National Academy of Sciences, the National Academy of Engineering, and the Institute of Medicine. The members of the committee responsible for the report were chosen for their special competences and with regard for appropriate balance.

Support for this project was provided by Contract DE-DT0001679 between the National Academy of Sciences and the Department of Energy and the National Nuclear Security Administration. Any opinions, findings, conclusions, or recommendations expressed in this publication are those of the authors and do not necessarily reflect the view of the agency that provided support for the project.

Cover: Photo of an OMEGA cryogenic implosion, courtesy of the University of Rochester's Laboratory for Laser Energetics.

International Standard Book Number-13: 978-0-309-27062-5
International Standard Book Number-10: 0-309-27062-6

Copies of this report are available free of charge from:

Board on Physics and Astronomy
National Research Council
The Keck Center of the National Academies
500 Fifth Street, NW
Washington, DC 20001

Additional copies of this report are available from the National Academies Press, 500 Fifth Street, NW, Keck 360, Washington, DC 20001; (800) 624-6242 or (202) 334-3313; http://www.nap.edu.

Printed in the United States of America

THE NATIONAL ACADEMIES

Advisers to the Nation on Science, Engineering, and Medicine

The National Academy of Sciences is a private, nonprofit, self-perpetuating society of distinguished scholars engaged in scientific and engineering research, dedicated to the furtherance of science and technology and to their use for the general welfare. Upon the authority of the charter granted to it by the Congress in 1863, the Academy has a mandate that requires it to advise the federal government on scientific and technical matters. Dr. Ralph J. Cicerone is president of the National Academy of Sciences.

The National Academy of Engineering was established in 1964, under the charter of the National Academy of Sciences, as a parallel organization of outstanding engineers. It is autonomous in its administration and in the selection of its members, sharing with the National Academy of Sciences the responsibility for advising the federal government. The National Academy of Engineering also sponsors engineering programs aimed at meeting national needs, encourages education and research, and recognizes the superior achievements of engineers. Dr. Charles M. Vest is president of the National Academy of Engineering.

The Institute of Medicine was established in 1970 by the National Academy of Sciences to secure the services of eminent members of appropriate professions in the examination of policy matters pertaining to the health of the public. The Institute acts under the responsibility given to the National Academy of Sciences by its congressional charter to be an adviser to the federal government and, upon its own initiative, to identify issues of medical care, research, and education. Dr. Harvey V. Fineberg is president of the Institute of Medicine.

The National Research Council was organized by the National Academy of Sciences in 1916 to associate the broad community of science and technology with the Academy's purposes of furthering knowledge and advising the federal government. Functioning in accordance with general policies determined by the Academy, the Council has become the principal operating agency of both the National Academy of Sciences and the National Academy of Engineering in providing services to the government, the public, and the scientific and engineering communities. The Council is administered jointly by both Academies and the Institute of Medicine. Dr. Ralph J. Cicerone and Dr. Charles M. Vest are chair and vice chair, respectively, of the National Research Council.

www.national-academies.org

**PANEL ON THE ASSESSMENT OF
INERTIAL CONFINEMENT FUSION TARGETS**

JOHN AHEARNE, NAE, Sigma Xi, *Chair*
DOUGLAS EARDLEY, University of California, Santa Barbara, *Vice Chair*
ROBERT DYNES, University of California, Santa Barbara
DAVID HARDING, University of Rochester
THOMAS MEHLHORN, Naval Research Laboratory
MERRI WOOD-SCHULTZ, Los Alamos, New Mexico
GEORGE ZIMMERMAN, Lawrence Livermore National Laboratory

Staff

GREG EYRING, Study Director
SARAH CASE, Study Director (until October 2011)
LaNITA JONES, Administrative Coordinator

BOARD ON PHYSICS AND ASTRONOMY

Preface and Acknowledgments

In the fall of 2010, the Office of the U.S. Department of Energy's (DOE's) Under Secretary for Science asked for a National Research Council (NRC) committee to investigate the prospects for generating power using inertial confinement fusion (ICF) concepts, acknowledging that a key test of viability for this concept—ignition[1]—could be demonstrated at the National Ignition Facility (NIF) at Lawrence Livermore National Laboratory (LLNL) in the relatively near term. The committee was asked to provide an unclassified report. However, DOE indicated that to fully assess this topic, the committee's deliberations would have to be informed by the results of some classified experiments and information, particularly in the area of ICF targets and nonproliferation. Thus, an additional Panel on Fusion Target Physics ("the panel") was assembled, composed of experts able to access the needed information (for member biographies, see Appendix A). The panel was charged with advising the committee on these issues, both by internal discussion and by this unclassified report. The statement of task for the panel is as follows:

> A Panel on Fusion Target Physics ("the panel") will serve as a technical resource to the Committee on Inertial Confinement Energy Systems ("the Committee") and will prepare a report that describes the R&D challenges to providing suitable targets, on the basis of parameters established and provided to the Panel by the Committee.

---

[1] The operative definition of ignition adopted by the panel, "gain greater than unity," is the same as that used in the earlier NRC report *Review of the Department of Energy's Inertial Confinement Fusion Program,* Washington, D.C.: National Academy Press (1997).

> The Panel on Fusion Target Physics will prepare a report that will assess the current performance of fusion targets associated with various ICF concepts in order to understand:
> 1. The spectrum output;
> 2. The illumination geometry;
> 3. The high-gain geometry; and
> 4. The robustness of the target design.
>
> The panel will also address the potential impacts of the use and development of current concepts for Inertial Fusion Energy on the proliferation of nuclear weapons information and technology, as appropriate. The Panel will examine technology options, but will not provide recommendations specific to any currently operating or proposed ICF facility.

The panel interpreted the terms used in its statement of task in the following way. "Illumination geometry" not only is interpreted to mean the physical arrangement and timing of laser or particle beams incident on the target but also is generalized to mean "delivering driver energy to the target." In this way, the magnetic forces in pulsed-power schemes are also included. "High-gain geometry" is interpreted as designs that enable the energy incident on the target to be converted efficiently into fuel burn and high yield.[2] "Spectrum output" is interpreted to include all of the types of emissions (photons, ions, neutrons, and debris) from the fusion target and their energy spectra. Depending on the type of reaction chamber used (solid wall, wetted wall, liquid wall, gas-filled, evacuated, and so on) these emissions may or may not reach the chamber wall; however, a detailed discussion of the effects on the wall is beyond the scope of this report. "Robustness of the target design" is interpreted in two ways: (1) the inherent "physics robustness," which relates to the performance margins of the design being large enough compared to the physics uncertainties that reliable performance can be assured under ideal conditions, and (2) "engineering robustness," which relates to the target's ability to deliver reliable performance even under nonideal conditions such as variations in driver energy, target manufacturing defects, errors in target positioning, or driver beam misalignment.

This unclassified report contains all of the panel's conclusions and recommendations. In some cases, additional support and documentation required the discussion of classified material, which appears in classified appendices in a separate version of this report. ICF is an active research field, and scientific understanding continues to evolve. The information discussed here is accurate as of the date presented to the panel (see Appendix B), although in some cases more recent updates are included; if so, this is noted in the text.

This report was reviewed in draft form by individuals chosen for their diverse perspectives and technical expertise in accordance with procedures approved by

[2] High yield is defined broadly as much more than 10 times the fusion energy produced as driver energy delivered to the target.

the Report Review Committee of the National Research Council. The purpose of this independent review is to provide candid and critical comments that will assist the institution in making its published report as sound as possible and to ensure that the report meets institutional standards for objectivity, evidence, and responsiveness to the study charge. The review comments and draft manuscript remain confidential to protect the integrity of the deliberative process.

We wish to thank the following individuals for their review of this report:

Bedros Afeyan, Polymath Research Inc.,
Roger Bangerter, E.O. Lawrence Berkeley National Laboratory (retired),
Michael Corradini, University of Wisconsin,
Jill Dahlburg, Naval Research Laboratory,
Richard Garwin, IBM Thomas J. Watson Research Center,
David Hammer, Cornell University,
Frank von Hippel, Princeton University,
Arjun Makhijani, Institute for Energy and Environmental Research,
David Overskei, Decision Factors Inc.,
Robert Rosner, University of Chicago, and
Douglas Wilson, Los Alamos National Laboratory.

Although the reviewers listed above have provided many constructive comments and suggestions, they were not asked to endorse the conclusions or recommendations, nor did they see the final draft of the report before its release. The review of this report was overseen by Louis J. Lanzerotti, New Jersey Institute of Technology. Appointed by the National Research Council, he was responsible for making certain that an independent examination of this report was carried out in accordance with institutional procedures and that all review comments were carefully considered. Responsibility for the final content of this report rests entirely with the authoring committee and the institution.

The panel also thanks the NRC staff for its dedicated work, in particular Sarah Case, who got the panel started off on the correct path, and Greg Eyring, who persevered in getting both the classified and unclassified reports over many hurdles.

John F. Ahearne, *Chair*
Panel on the Assessment of Inertial Confinement Fusion Targets

Contents

Summary

In the fall of 2010, the Office of the U.S. Department of Energy's (DOE's) Under Secretary for Science asked for a National Research Council (NRC) committee to investigate the prospects for generating power using inertial fusion energy (IFE), noting that a key test of viability for this concept—ignition[1]—could be demonstrated at the National Ignition Facility (NIF) at Lawrence Livermore National Laboratory (LLNL) in the relatively near term. In response, the NRC formed both the Committee on the Assessment of the Prospects for Inertial Fusion Energy ("the committee") to investigate the overall prospects for IFE in an unclassified report and the separate Panel on Fusion Target Physics ("the panel") to focus on issues specific to fusion targets, including the results of relevant classified experiments and classified information on the implications of IFE targets for the proliferation of nuclear weapons.

This is the report of the Panel on Fusion Target Physics, which is intended to feed into the broader assessment of IFE being done by the NRC committee. It consists of an unclassified body, which contains all of the panel's conclusions and recommendations, as well as three classified appendices, which provide additional support and documentation.

[1] The operative definition of ignition adopted by the panel, "gain greater than unity," is the same as that used in the earlier NRC report *Review of the Department of Energy's Inertial Confinement Fusion Program*, Washington, D.C.: National Academy Press (1997).

BACKGROUND

Fusion is the process by which energy is produced in the sun, and, on a more human scale, is the one of the key processes involved in the detonation of a thermonuclear bomb. If this process could be "tamed" to provide a controllable source of energy that can be converted to electricity—as nuclear fission has been in currently operating nuclear reactors—it is possible that nuclear fusion could provide a new method for producing low-carbon electricity to meet U.S. and world growing energy needs.

For inertial fusion to occur in a laboratory, fuel material (typically deuterium and tritium) must be confined for an adequate length of time at an appropriate density and temperature to overcome the Coulomb repulsion of the nuclei and allow them to fuse. In inertial confinement fusion (ICF)—the concept investigated in this report[2]—a driver (e.g., a laser, particle beam, or pulsed magnetic field) delivers energy to the fuel target, heating and compressing it to the conditions required for ignition. Most ICF concepts compress a small amount of fuel directly to thermonuclear burn conditions (a hot spot) and propagate the burn via alpha particle deposition through adjacent high-density fuel regions, thereby generating a significant energy output.

There are two major concepts for inertial confinement fusion target design: direct-drive targets, in which the driver energy strikes directly on the fuel capsule, and indirect-drive targets, in which the driver energy first strikes the inside surface of a hollow chamber (a hohlraum) surrounding the fuel capsule, producing energetic X-rays that compress the fuel capsule. Conventional direct and indirect drive share many key physics issues (e.g., energy coupling, the need for driver uniformity, and hydrodynamic instabilities); however, there are also issues that are unique to each concept.

The only facility in the world that was designed to conduct ICF experiments that address the ignition scale is the NIF at LLNL. The NIF driver is a solid-state laser. For the first ignition experiments, the NIF team has chosen indirect-drive targets. The NIF can also be configured for direct drive. In addition, important work on laser-driven, direct-drive targets (albeit at less than ignition scale) is also under way in the United States at the Naval Research Laboratory and the OMEGA laser at the University of Rochester. Heavy-ion-beam drivers are being investigated at the Lawrence Berkeley National Laboratory (LBNL), LLNL, and the Princeton Plasma Physics Laboratory (PPPL), and magnetic implosion techniques are being explored on the Z machine at Sandia National Laboratories (SNL) and at Los

[2] Inertial confinement fusion (ICF) is the process by which the target is heated and compressed by the driver to reach fusion conditions. Inertial fusion energy (IFE) is the process by which useful energy is extracted from ignition and burn of ICF fuel targets.

Alamos National Laboratory (LANL). Important ICF research is also under way in other countries, as discussed later in this report.

SPECIFIC CONCLUSIONS AND RECOMMENDATIONS

The panel's key conclusions and recommendations, all of them specific to various aspects of inertial confinement fusion, are presented below. They are labeled according to the chapter and number order in which they appear in the text, to provide the reader with an indicator of where to find a more complete discussion. This summary ends with two overarching conclusions and an overarching recommendation derived from viewing all of the information presented to the panel as a whole.

Targets for Indirect Laser Drive

CONCLUSION 4-1: The national program to achieve ignition using indirect laser drive has several physics issues that must be resolved if it is to achieve ignition. At the time of this writing, the capsule/hohlraum performance in the experimental program, which is carried out at the NIF, has not achieved the compressions and neutron yields expected based on computer simulations. At present, these disparities are not well understood. While a number of hypotheses concerning the origins of the disparities have been put forth, it is apparent to the panel that the treatments of the detrimental effects of laser-plasma interactions (LPI) in the target performance predictions are poorly validated and may be very inadequate. A much better understanding of LPI will be required of the ICF community.

CONCLUSION 4-2: Based on its analysis of the gaps in current understanding of target physics and the remaining disparities between simulations and experimental results, the panel assesses that ignition using laser indirect drive is not likely in the next several years.

The National Ignition Campaign (NIC) plan—as the panel understands it—suggests that ignition is planned after the completion of a tuning program lasting 1-2 years that is presently under way and scheduled to conclude at the end of FY2012. While this success-oriented schedule remains possible, resolving the present issues and addressing any new challenges that might arise are likely to push the timetable for ignition to 2013-2014 or beyond.

Targets for Indirect-Drive Laser Inertial Fusion Energy

CONCLUSION 4-4: The target design for a proposed indirect-drive IFE system (the Laser Inertial Fusion Energy, or LIFE, program developed by LLNL) incorporates

plausible solutions to many technical problems, but the panel assesses that the robustness of the physics design for the LIFE target concept is low.

- The proposed LIFE target presented to the panel has several modifications relative to the target currently used in the NIC (e.g., rugby hohlraums, shine shields, and high-density carbon ablators), and the effects of these modifications may not be trivial. For this reason, R&D and validation steps would still be needed.
- There is no evidence to indicate that the margin in the calculated target gain ensures either its ignition or sufficient gain for the LIFE target. If ignition is assumed, then the gain margin briefed to the panel, which ranged from 25 percent to almost 60 percent when based on a calculation that used hohlraum and fuel materials characteristic of the NIC rather than the LIFE target, is unlikely to compensate for the phenomena relegated to it—for example, the effects of mix—under any but the most extremely favorable eventuality. In addition, the tight coupling of LIFE to what can be tested on the NIF constrains the potential design space for laser-driven, indirect-drive IFE.

Targets for Direct-Drive Laser Inertial Fusion Energy

CONCLUSION 4-6: The prospects for ignition using laser direct drive have improved enough that it is now a plausible alternative to laser indirect drive for achieving ignition and for generating energy.

- The major concern with laser direct drive has been the difficulty of achieving the symmetry required to drive such targets. Advances in beam-smoothing and pulse-shaping appear to have lessened the risks of asymmetries. This assessment is supported by data from capsule implosions (performed at the University of Rochester's OMEGA laser), but it is limited by the relatively low drive energy of the implosion experiments that have thus far been possible. Because of this, the panel's assessment of laser-driven, direct-drive targets is not qualitatively equivalent to that of laser-driven, indirect-drive targets.
- Further evaluation of the potential of laser direct-drive targets for IFE will require experiments at drive energies much closer to the ignition scale.
- Capsule implosions on OMEGA have established an initial scaling point that indicates the potential of direct-drive laser targets for ignition and high yield.
- Polar direct-drive targets[3] will require testing on the NIF.

[3] In polar direct drive, the driver beams are clustered in one or two rings at opposing poles. To increase the uniformity of the drive, polar drive beams strike the capsule obliquely, and the driver energy is biased in favor of the more equatorial beams.

- Demonstration of polar-drive ignition on the NIF will be an important step toward an IFE program.
- If a program existed to reconfigure the NIF for polar drive, direct-drive experiments that address the ignition scale could be performed as early as 2017.

Fast Ignition

Fast ignition (FI) requires a combination of long-pulse (implosion) and short-pulse (ignition) lasers. Aspects of fast ignition by both electrons and protons were briefed to the panel. Continued fundamental research into fast ignition theory and experiments, the acceleration of electrons and ions by ultrashort-pulse lasers, and related high-intensity laser science is justified. However, issues surrounding low laser-target energy coupling, a complicated target design, and the existence of more promising concepts (such as shock ignition) led the panel to the next conclusion regarding the relative priority of fast ignition for fusion energy.

CONCLUSION 4-5: At this time, fast ignition appears to be a less promising approach for IFE than other ignition concepts.

Laser-Plasma Interactions

A variety of LPI take place when an intense laser pulse hits the target capsule or surrounding hohlraum. Undesirable effects include backscattering of laser light, which can result in loss of energy; cross-beam energy transfer among intersecting laser beams, which can cause loss of energy or affect implosion symmetry; acceleration of suprathermal "hot electrons," which then can penetrate and preheat the capsule's interior and limit later implosion; and filamentation, a self-focusing instability that can exacerbate other LPI. LPI have been a key limiting factor in laser inertial confinement fusion, including the NIC indirect-drive targets, and are still incompletely understood.

CONCLUSION 4-11: The lack of understanding surrounding laser-plasma interactions remains a substantial but as yet unquantified consideration in ICF and IFE target design.

RECOMMENDATION 4-1: DOE should foster collaboration among different research groups on the modeling and simulation of laser-plasma interactions.

Heavy-Ion Targets

A wide variety of heavy-ion target designs has been investigated, including indirect-drive, hohlraum/capsule targets that resemble NIC targets. Recently, the emphasis has shifted to direct-drive targets, but to date the analysis of how these targets perform has been based on computation rather than experiment, and the codes have not been benchmarked with experiments in relevant regimes.

CONCLUSION 4-12: The U.S. heavy-ion-driven fusion program is considering direct-drive and indirect-drive target concepts. There is also significant current work on advanced target designs.[4] This work is at a very early stage, but if successful may provide very high gain.

- The work in the heavy-ion fusion (HIF) program involves solid and promising science.
- Work on heavy-ion drivers is complementary to the laser approaches to IFE and offers a long-term driver option for beam-driven targets.
- The HIF program relating to advanced target designs is in a very early stage and is unlikely to be ready for technical assessment in the near term.
- The development of driver technology will take several years, and the cost to build a significant accelerator driver facility for any target is likely to be very high.

Z-Pinch Targets

Current Z-pinch direct-drive concepts utilize the pressure of a pulsed, high magnetic field to implode deuterium-tritium fuel to fusion conditions. Simulations predict that directly using the pressure of the magnetic field to implode and compress the target can greatly increase the efficiency with which the electrical energy is coupled to the fuel as compared with the efficiency of indirect drive from Z-pinch X-ray sources. There is work under way on both classified and unclassified target designs.

CONCLUSION 4-13: Sandia National Laboratories is leading a research effort on a Z-pinch scheme that has the potential to produce high gain with good energy efficiency, but concepts for an energy delivery system based on this driver are too immature to be evaluated at this time.

[4] Advanced designs include direct-drive, conical X-target configurations (see Chapter 2).

It is not yet clear that the work at SNL will ultimately result in the high gain predicted by computer simulations, but initial results are promising and it is the panel's opinion that significant progress in the physics may be made in a year's time. The pulsed-power approach is unique in that its goal is to deliver a large amount of energy (~10 MJ) to targets with good efficiency (≥10 percent) and to generate large fusion yields at low repetition rates.

Target Fabrication

Current targets for inertial confinement fusion experiments tend to be one-off designs, with specifications that change according to the experiments being run. In contrast, targets for future IFE power plants will have to have standard, low-cost designs that are mass-produced in numbers as high as a million targets per day per power plant. The panel examined the technical feasibility of producing targets for various drivers, including limited aspects of fabrication for IFE. However, a full examination of the issues of mass production and low cost is the province of the NRC IFE committee study.

CONCLUSION 4-7: In general, the science and engineering of manufacturing fusion targets for laser-based ICF are well advanced and meet the needs of those experiments, although additional technologies may be needed for IFE. Extrapolating this status to predict the success of manufacturing IFE targets is reasonable if the target is only slightly larger than the ICF target and the process is scalable. However, subtle additions to the design of the ICF target to improve its performance (greater yield) and survivability in an IFE power plant may significantly affect the manufacturing paradigm.

Proliferation Risks of IFE

Many modern nuclear weapons rely on a fusion stage as well as a fission stage, and there has been discussion of the potential for host state proliferation—particularly vertical proliferation—associated with the siting of an IFE power plant. The panel was asked to evaluate the proliferation risks associated with IFE, particularly with regard to IFE targets.

CONCLUSION 3-1: At present, there are more proliferation concerns associated with indirect-drive targets than with direct-drive targets. However, the spread of technology around the world may eventually render these concerns moot. Remaining concerns are likely to focus on the use of classified codes for target design.

CONCLUSION 3-2: The nuclear weapons proliferation risks associated with fusion power plants are real but are likely to be controllable. These risks fall into three categories:

- **Knowledge transfer,**
- **Special nuclear material (SNM) production, and**
- **Tritium diversion.**

OVERARCHING CONCLUSIONS AND RECOMMENDATION

While the focus of this panel was on ICF target physics, the need to evaluate driver-target interactions required considering driver characteristics as well. This broader analysis led the panel to the following overarching conclusions and a recommendation.

OVERARCHING CONCLUSION 1: The NIF has the potential to support the development and further validation of physics and engineering models relevant to several IFE concepts, from indirect-drive hohlraum designs to polar direct-drive ICF and shock ignition.

- **In the near to intermediate term, the NIF is the only platform that can provide information relevant to a wide range of IFE concepts at ignition scale. Insofar as target physics is concerned, it is a modest step from NIF scale to IFE scale.**
- **Targets for all laser-driven IFE concepts (both direct-drive and indirect-drive) can be tested on the NIF. In particular, reliable target performance would need to be demonstrated before investments could confidently be made in the development of laser-driven IFE target designs.**

The NIF will also be helpful in evaluating indirectly driven, heavy-ion targets. It will be less helpful in gathering information relevant to current Z-pinch, heavy-ion direct drive, and heavy-ion advanced target concepts.

OVERARCHING CONCLUSION 2: It would be advantageous to continue research on a range of IFE concepts, for two reasons:

- **The challenges involved in the current laser indirect-drive approach in the single-pulse National Nuclear Security Administration program at the NIF have not yet been resolved, and**
- **The alternatives to laser indirect drive have technical promise to produce high gain.**

In particular, the panel concludes that laser direct drive is a viable concept to be pursued on the NIF. SNL's work on Z-pinch can serve to mitigate risk should the NIF not operate as expected. This work is at a very early stage but is highly complementary to the NIF approach, because none of the work being done at SNL relies on successful ignition at the NIF, and key aspects of the target physics can be investigated on the existing Z-machine. Finally, emerging heavy-ion designs could be fruitful in the long term.

OVERARCHING RECOMMENDATION: The panel recommends against pursuing a down-select decision for IFE at this time, either for a specific concept such as LIFE or for a specific target type/driver combination.

Further R&D will be needed on indirect drive and other ICF concepts, even following successful ignition at the NIF, to determine the best path for IFE in the coming decades.

1

Introduction

Inertial fusion energy (IFE) has been a concept since the 1970s, and the National Research Council (NRC) has performed several reviews of the Department of Energy's (DOE's) programs for inertial confinement fusion (ICF)—the essential concept underlying IFE—since that time (NRC, 1986, 1990, and 1997). This report of the Panel on Fusion Target Physics supports and informs a broader study on the prospects for IFE being undertaken by a separate NRC committee.[1] The broader study is motivated by a desire on the part of DOE, the sponsor, to determine a clearer path forward for the IFE concept, in view of the prospect that a key test of viability for this concept—ignition—can be demonstrated at the National Ignition Facility (NIF) at Lawrence Livermore National Laboratory (LLNL) in the relatively near term.

To address its statement of task (see the Preface), the panel heard from many sources, listed in Appendix B, and visited several laboratories involved in U.S. efforts in ICF and IFE—LLNL, Sandia National Laboratories, Lawrence Berkeley National Laboratory, the University of Rochester Laboratory for Laser Energetics, and the Naval Research Laboratory—and heard from representatives of additional programs at the Los Alamos National Laboratory.

The panel's focus in this study is IFE targets, including both direct-drive and indirect-drive targets. To distinguish its role as clearly as possible from that of the main study committee, the panel drew a conceptual sphere around the outside of the target and considered anything crossing the surface of the sphere (energy

[1] The Committee on the Prospects for Inertial Confinement Fusion Energy Systems.

coming in, reaction products going out) as well as physics processes taking place inside the sphere, to be within its purview. In addition, the panel considered the technical feasibility of fabricating various target concepts to be within its charge, but deemed the mass manufacturing of high-performance, cost-effective targets for future power plants to be part of the main committee's responsibility. Inevitably, there were certain topics at the interface between the charges of the panel and the main committee, such as the survivability of the injected target in the extreme environment of the reaction chamber. In such cases, the panel felt that it was preferable that the panel and committee reports should overlap rather than risk the possibility that important topics might be left out.

Chapter 2 provides a brief technical background on IFE and a discussion of key concepts related to ICF targets and their role in IFE. In Chapter 3, the proliferation risks of specific target designs are discussed, as well as the broader proliferation risks associated with IFE plants and research facilities. Chapter 4 evaluates the current status of various targets, considering the results of actual experiments on their performance as well as the analytical and predictive capabilities of available codes and simulations. This analysis is used to characterize the state of our current understanding of fusion target physics and to identify the major issues that remain to be resolved. The classified version of this report contains additional appendixes discussing classified material that the panel considers relevant to its conclusions and recommendations.

2

Technical Background

This chapter briefly introduces the key concepts necessary to understand inertial confinement fusion (ICF), inertial fusion energy (IFE), and target physics.

INERTIAL CONFINEMENT FUSION AND INERTIAL FUSION ENERGY

Nuclear fusion—the process by which the nuclei of atoms such as deuterium or tritium combine to form a heavier nucleus, such as that of helium—can release a significant amount of energy. Fusion is the process by which energy is produced in the sun and, on a more human scale, is the one of the key processes involved in the detonation of a thermonuclear bomb.

If this process can be tamed to provide a controllable source of energy that can be converted to electricity—as the nuclear fission process is used in nuclear reactors—then it is possible that nuclear fusion could be a new way to produce low-carbon electricity to meet the growing energy needs of the United States and the world. However, this possibility is far from imminent, and a great deal of scientific and engineering work remains to be done before a commercial nuclear fusion plant can be demonstrated.

For inertial fusion to occur in a laboratory, heating of the fuel material (typically deuterium and tritium) must be confined to a small enough hot spot to overcome the Coulomb repulsion of the nuclei and allow fusion to initiate in a small region of the fuel ("ignition"). If successful, this process will release sufficient energy to sustain the fusion "burn" that will propagate through the fuel, generating a significant energy output. Two concepts are typically discussed for accomplishing this confinement: (1) magnetic confinement fusion (MCF), in which magnetic fields are used to confine the plasma, and (2) ICF, the topic of the current report, in which a driver delivers energy to the surface of a pellet of fuel, heating and compressing it. Potential drivers include lasers, particle beams, and X-rays, among other concepts.

In ICF, energy supplied by the driver is applied, either directly or indirectly, to the outer layer of a fuel pellet that is typically made up of an ablator material (e.g., beryllium, doped plastic, or high-density carbon) that explodes outward as it heats. This outward explosion of the surface layer forces the remainder of the fuel (typically light elements such as deuterium and tritium) to accelerate inward to conserve momentum. The timing of the inward fuel acceleration is controlled carefully in order to compress the fuel using a minimum of energy. At the same time, sudden increases in the driver power profile both accelerate the implosion and send shock waves into the center of the fuel, heating it sufficiently that fusion reactions begin to occur.[1]

The goal of ICF is to initiate a self-sustaining process in which the energetic alpha particles emitted by the ongoing fusion reactions heat the surrounding fuel to the point where it also begins to undergo fusion reactions. The percentage of fuel that undergoes fusion is referred to as the "burn-up fraction." The fuel gain G (defined as the ratio of the total energy released by the target to the driving beam energy impinging upon it) depends on the burn-up fraction, and gains greater than about 10 will need to be demonstrated to validate the target physics of any approach to a practical IFE power plant.

Important target physics includes processes that deflect or absorb driver energy within the target; the transport of energy within the target; capsule preheat; conversion of energy to the inward-directed implosion by ablation; fuel compression and heating; thermonuclear reactions; transport and deposition of neutron and alpha-particle energy, resulting in bootstrapping thermonuclear reactions; and hydrodynamic disassembly and output. Models exist for all of these processes, but some are more predictive than others. Some processes are difficult to simulate, such as laser-plasma interactions, the generation and transport of hot electrons in self-consistent magnetic fields, nonlocal-thermal-equilibrium atomic physics, hydrodynamic instabilities, mix, and debris generation. These models continue to evolve to keep pace with experiments. Other processes, such as large-scale hydrodynamics, thermonuclear reactions, and X-ray-, neutron- and alpha-particle transport appear to be simulated adequately using standard numerical models.

The Department of Energy (DOE) is funding multiple efforts to investigate the physics of ICF; many of these efforts have the potential to inform current

[1] What is described here is known as hot-spot ignition; other potential concepts for ignition are being considered and are introduced briefly later in this chapter.

understanding of the prospects for IFE. Over the next several years, experiments will be ongoing at the National Ignition Facility (NIF) at Lawrence Livermore National Laboratory (LLNL) that are aimed at achieving ICF ignition. At the same time, experiments such as those at the University of Rochester's Laboratory for Laser Energetics, the Naval Research Laboratory, Lawrence Berkeley Laboratory, and Sandia National Laboratories continue to advance our understanding and control of ICF using different technology and physics approaches. However, it should be recognized that up to this point, the majority of the funding and efforts related to ICF target physics are provided by—and related to—the U.S. nuclear weapons program and its stockpile stewardship efforts and are not directly aimed at energy applications.

DOE's Centurion-Halite program revolved around a series of underground experiments conducted in the 1980s in which target capsules were driven by the energy from nuclear explosions. Additional discussion of the program is provided in classified Appendix D.

BASICS OF ICF TARGET PHYSICS AND DESIGN

Target Design: Direct and Indirect Drive, Z-Pinch

There are two major concepts for ICF target design: direct-drive targets, in which the driver energy (e.g., in the form of laser beams, particle beams, or magnetic field pressure) directly strikes the fuel capsule (see Figure 2-1); and indirect-drive targets, in which the driver energy first strikes a hollow chamber (a "hohlraum") surrounding the fuel capsule, producing energetic X-rays that compress the fuel capsule (see Figure 2-2). Conventional direct and indirect drive share many key physics issues, such as energy coupling, the need for driver uniformity, and hydrodynamic instabilities; however, there are issues that are unique to each concept.

Generally, the elements of the fuel capsule are similar for direct drive and indirect drive, at least with respect to laser drivers. Fuel capsules are typically spherical, with several layers: an outer ablator layer; a layer of cryogenic frozen fuel; and a center of gaseous fuel, typically deuterium-tritium (DT). A sample fuel capsule is shown in Figure 2-3.

Several of the key differences between direct drive and indirect drive for ICF are discussed briefly in the sections that follow.

Direct Drive

Direct-drive concepts for ICF using laser drivers are currently being researched at the University of Rochester's Laboratory for Laser Energetics (LLE) and the Naval Research Laboratory (NRL). Concepts using heavy-ion beam drivers are being

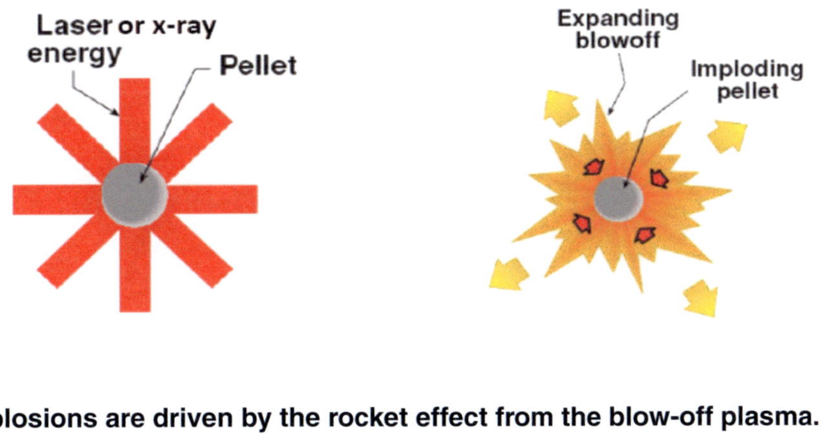

FIGURE 2-1 In the case of direct drive, the fuel pellet is illuminated symmetrically by the driver energy, resulting in implosion. SOURCE: R. Betti, University of Rochester, "Tutorial on the Physics of Inertial Confinement Fusion," presentation to the NRC IFE committee on April 22, 2011.

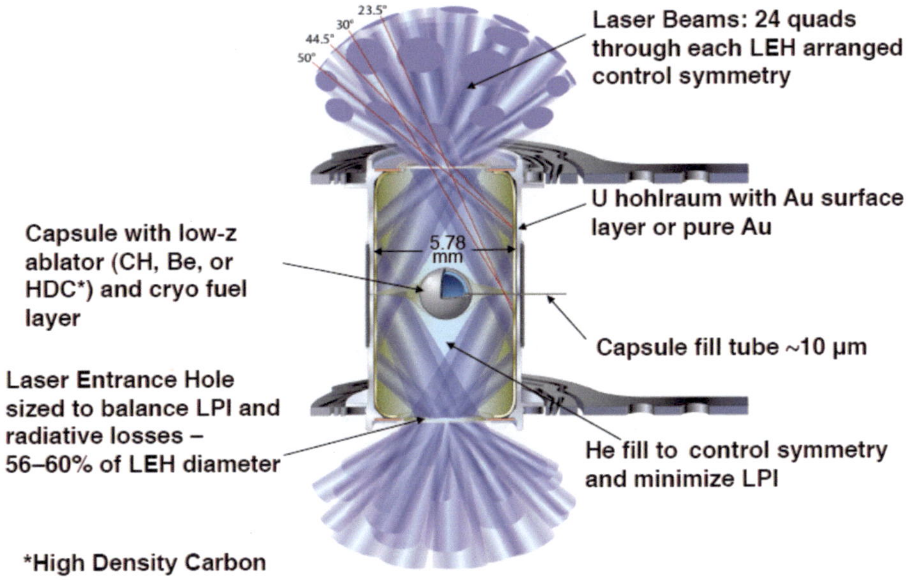

FIGURE 2-2 In the case of indirect drive, driver energy incident on a hohlraum is converted to X-rays, which then impinge symmetrically on the fuel capsule, causing it to implode. This figure shows the laser beam geometry used in the National Ignition Campaign (NIC) at LLNL. LEH, laser entrance hole; LPI, laser-plasma interactions; HDC, high-density carbon. SOURCE: J. Lindl, LLNL, "The National Ignition Campaign on NIF and Its Extension to Targets for IFE," presentation to the panel on February 16, 2011.

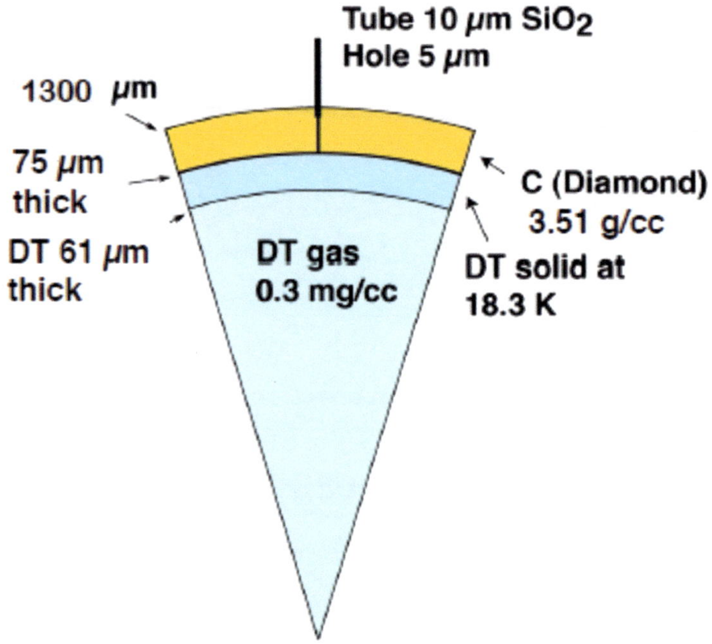

FIGURE 2-3 Section of a spherical fuel capsule design showing the ablator layer (in this case pure carbon), a layer of DT ice, and an inner core of DT gas. SOURCE: J. Lindl, LLNL, "The National Ignition Campaign on NIF and Its Extension to Targets for IFE," presentation to the panel on February 16, 2011.

studied at Lawrence Berkeley National Laboratory (LBNL), and Sandia National Laboratories (SNL) is developing direct-drive concepts for pulsed-power drivers.

The major benefit of direct-drive target design is the calculated potential for higher energy gain than indirect drive. This relatively large gain is in large part due to avoiding the losses that occur during the conversion of laser beams or particle beams to X-rays in the hohlraum, discussed in detail in the next section. Avoiding these losses results in a higher percentage of driver energy absorbed by the capsule in direct drive, increasing the efficiency and potentially decreasing the size of the driver required.

Polar direct drive is a variant of the spherically symmetric, direct-drive illumination geometry shown in Figure 2-1. As shown in Figure 2-4, the driver beams are clustered in one or two rings at opposing poles. To increase the uniformity of the drive, polar drive beams strike the capsule obliquely, and the driver energy is biased in favor of the more equatorial beams. Although the polar illumination geometry is consequently less efficient than the spherically symmetric geometry, it is more compatible with the current NIF configuration.

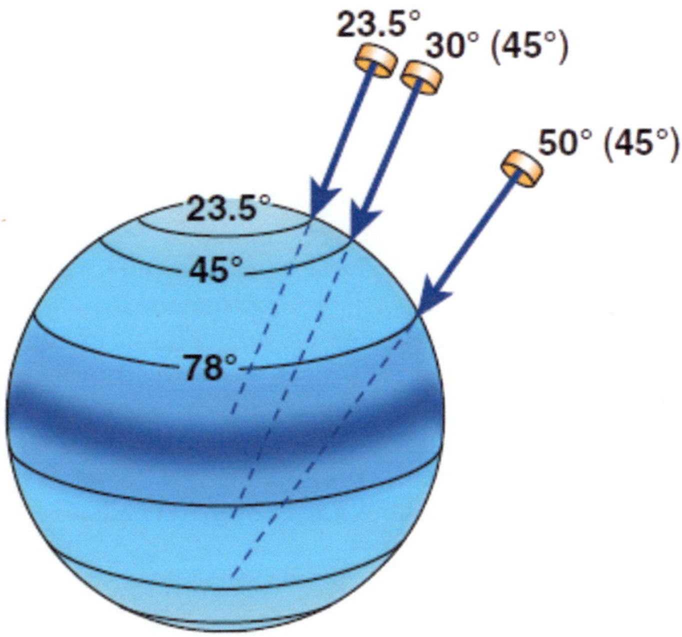

FIGURE 2-4 In the polar direct-drive illumination geometry, the driver beams are incident from directions above and below the fuel capsule but not near the equator. SOURCE: R. L. McCrory, University of Rochester, "Laser-Driven Inertial Fusion Energy: Direct-Drive Targets Overview," presentation to the panel on February 16, 2011.

Since the 1980s, there has been an ongoing effort in laser science that has focused on improving the performance of direct-drive laser systems for both solid-state and KrF lasers. For solid-state lasers, these advances include frequency tripling (for improved energy coupling and lower instability growth rates), smoothing by spectral dispersion (SSD), and polarization smoothing, to reduce imprinting of beam nonuniformities on the target. Recently LLE developed SSD with multiple phase-modulation frequencies (Multi-FM) and proposed using this technique to modify the NIF for polar direct drive.

High-energy KrF lasers were developed to utilize the deep ultraviolet (248 nm) wavelength of the system. Induced spatial incoherence (ISI) was developed to smooth the beams, and recently focal zooming[2] was demonstrated to improve the efficiency of coupling the laser with imploding targets. Direct-drive target experiments on the OMEGA laser have shown steady improvement toward theoretical

[2] Zooming involves reducing the driver spot size to match the diameter of the imploding capsule, thereby increasing the efficiency of energy coupling between driver and target.

yield limits by combining a large number (60) of laser beams, better laser beam smoothing techniques, and improved beam pointing and target placement at the target chamber center. Although historically much of the discussion of direct-drive fusion has involved laser drivers (e.g., LLE's work at the OMEGA laser facility and the Nike KrF laser experiments at NRL), direct-drive ICF has potential for use with other drivers. In particular, the panel was briefed on direct-drive targets by members of the LBNL heavy-ion driver program.

However, there are difficulties involved in using direct-drive fusion. A direct-drive capsule must tolerate four major sources of perturbations to ignite and burn: drive asymmetry, inhomogeneous capsule surface finish, ice roughness in the layer between the cryogenic DT and the DT gas; and driver imprint.[3] The effects of the driver imprint and drive asymmetry are reduced for indirect drive. In addition, without a hohlraum to protect the capsule from the high temperatures in the chamber, and if there is no buffer gas to protect the chamber walls from emitted alpha particles, alternative methods must be found to address these threats.

Indirect Drive

As shown in Figure 2-2, indirect drive (whether using laser drivers or an alternative driver, such as heavy-ion beams) consists of driver beams entering a hohlraum, which is essentially a hollow cylinder, typically made of gold, or oblong capsule with (in the case of laser drivers) openings on either end. LLNL is currently leading research into indirect-drive concepts for laser-driven ICF at the NIF. The driver beams are directed to enter the openings on either end of the hohlraum and strike the interior of the hohlraum in four circular arrays, two near the center, and two nearer the ends (see Figure 2-2). The energy deposited by the laser beams on the interior of the hohlraum produces a hot plasma that radiates primarily in X-rays at a temperature of about 300 eV, or 3.3 million K. These X-rays are then absorbed by the capsule, resulting in implosion.

A virtue of the hohlraum in an actual IFE target is that it functions as a thermal shroud to protect the integrity of the cryogenic fuel capsule inside the target. This allows the target chamber to contain an inert gas (xenon) at low pressure to help protect the walls of the target chamber from X-rays emitted by high-Z materials in the exploding target.

---

[3] For laser drivers, driver imprint occurs early in time when the target ablator is cold and dense. It is related to the asymmetries from modulations in individual laser beams (short wavelength) and perturbations from overlapping drive beams or by beams with slightly differing arrival times and angles of incidence (longer wavelength).

Benefits of Indirect Drive for Smoothing

Spatial nonuniformities at any scale can significantly increase the deviation of the actual implosion of an inertial fusion capsule from the conditions it was designed to achieve, with the result that the conditions inside the imploded capsule lie in a less favorable location in thermodynamic phase space than intended. Indirect drive of laser targets was conceived and developed to eliminate the effects of nonuniformities within each laser beam delivered to the target chamber.

The smoothing obtained through the use of indirect drive is a consequence of transforming the energy of each laser from a focused beam into thermal radiation. Any nonuniformity in a laser beam entering an indirect-drive target chamber transfers to the wall of the hohlraum enclosing the target, heating its material to a heterogeneous plasma. This heterogeneity is somewhat smoothed by energy transport processes within the radiating plasma itself, but a stronger smoothing effect occurs because the X-rays originating in each localized mass of plasma affect the entire portion of the target capsule surface to which it has a direct line of sight. The result is that localized variations in X-ray emission are averaged over the capsule surface, and rapid changes of drive conditions over the surface of the capsule are eliminated.

The development and use of indirect drive was the primary focus of LLNL on the 10-beam NOVA laser. This experience led to the development of the NIF indirect-drive configuration, which is much more sophisticated, using 192 laser beams in inner and outer clusters to control symmetry and pulse shape (see Figure 2-2).

Although the capsule absorption of X-rays is more efficient than the direct absorption of laser light in direct-drive fusion, enough energy is lost in the heating of the hohlraum to significantly reduce the efficiency of indirect-drive fusion relative to direct-drive fusion. This results in lower calculated potential gains for indirect-drive fusion targets.

As with direct drive, although its primary development historically has been with laser drivers, indirect drive has been used in IFE system designs with other drivers (e.g., heavy ions and early Z-pinch schemes). The key is to deposit enough energy on the inner surface of the hohlraum to produce a hot plasma that radiates thermal X-rays.

One of the key reasons that indirect-drive targets were developed is that ICF can model on a laboratory scale some aspects of a thermonuclear explosion. This is highly useful for the applications of ICF at the NIF at LLNL that are related to the long-term stewardship of the U.S. nuclear stockpile. This motivation has been a key aspect in the development of the indirect-drive approach for IFE, since one could leverage insights from better-funded weapons programs for the less well funded energy programs. However, there remains debate about whether this provides significant benefits for energy generation using ICF, and some argue that the indirect-drive approach—if commercialized and distributed overseas—could increase the risk that nuclear weapons knowledge and information will proliferate. This topic is analyzed in more detail in the classified Appendix E and in Chapter 3.

Z-Pinch Target

In recent ICF and IFE studies, Z-pinch targets are imploded by the pressure of ultrahigh magnetic fields generated by high currents (e.g., 20-60 MA for ~100 ns) provided by pulsed-power generators rather than by the ablation pressure generated by illuminating a capsule with a high-power laser. While laser fusion capsules are typically spherical shells, Z-pinch targets are typically conducting cylindrical shells containing DT fuel. Because magnetic field strength increases inversely with the radius of the conductor in which the current flows (I/r), as long as the driver has the appropriate electrical characteristics to deliver current to the increasingly high-inductance target, the magnetic pressure (proportional to B^2) continues to grow, accelerating the cylindrical implosion and compressing the fuel. For appropriate design conditions, the DT fuel can be heated to sufficient temperature to initiate fusion reactions and compressed to sufficient areal density (bulk density ρ times fuel radius r) to trap emitted alpha particles and initiate bootstrap heating.

Physics of Different Types of Ignition

Hot-Spot Ignition

Hot-spot ignition, described briefly earlier in this chapter, is the most commonly discussed and best understood method for achieving ignition. Hot-spot ignition refers to the creation of a small central mass of fuel that is heated to temperatures sufficient to begin efficient thermonuclear burn (~10 keV), surrounded by a larger mass of dense but colder fuel that has sufficient areal density (>300 mg/cm^2) to trap alpha particles and initiate bootstrap heating.[4]

The primary reason for utilizing hot-spot ignition is to minimize the driver energy requirements. Heating fuel to 10 keV is energy-intensive, so the goal is to use the driver energy to launch a series of shocks that simultaneously coalesce and heat only a small central mass to fusion temperatures, while quasi-isentropically compressing the main fuel mass as close to the Fermi-degenerate limit (the minimum energy state for high-density matter) as possible. The energy deposited by fusion alpha particles rapidly heats the cold, dense main fuel, causing it to reach

[4] R.L. McCrory, University of Rochester, "Laser-Driven Inertial Fusion Energy: Direct-Drive Targets Overview," presentation to the panel on February 16, 2011.

thermonuclear burn conditions. The fusion burn terminates when the rapidly heated fuel mass overcomes the inertia of implosion and explodes to lower densities and temperatures where fusion reaction rates rapidly decrease (hence the term "inertial confinement").

In order to use minimum driver energy, it is important to compress most of the fuel near the Fermi-degenerate adiabat. At least four laser pulses are required to provide the compression energy in a time-dependent fashion that is consistent with this goal. More, smaller pulses—or even a continuous power profile—could also be used, but the four-pulse system is the easiest to control and observe experimentally.

Fast Ignition

In FI, ignition is separated from the compression phase. The fuel is compressed (using lasers or another driver) at a lower velocity than in hot-spot ignition. The goal is to create a fuel mass that has at least the 300 mg/cm^2 areal density required to capture alpha particles, but not the DT temperature to initiate fusion burn. The energy to ignite a small portion of this compressed fuel is provided by a high-intensity, ultrashort-pulse laser. For the correct conditions, the thermonuclear burn propagates from this heated fuel volume into the rest of the cold, imploded fuel.

The leading approach to fast ignition uses a hollow cone of high-density material inserted into the fuel capsule so as to allow clean entry of this second laser beam to the compressed fuel assembly (see Figure 2-5). The principle of fast ignition was first demonstrated at the Institute of Laser Engineering in Osaka, Japan, in experiments performed on the Gekko-XII laser (Kodama et al., 2002).

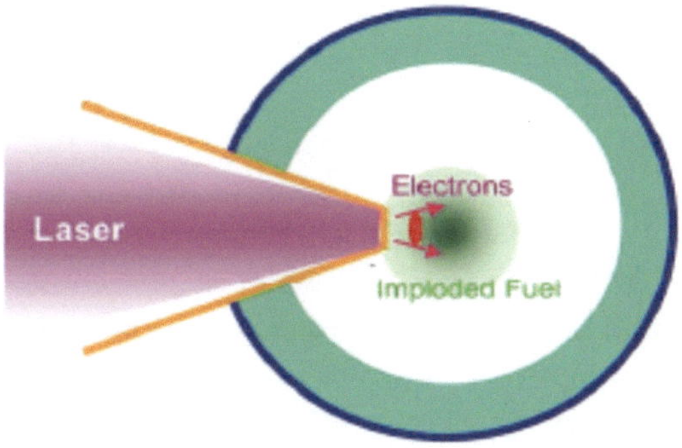

FIGURE 2-5 In this version of fast ignition, a short, high-intensity laser pulse enters the cone of a cone-and-capsule assembly after the fuel capsule has been compressed by an earlier pulse, producing a pulse of hot electrons that initiate fusion. SOURCE: Juan Fernandez, LANL, "Inertial Confinement Fusion Targets at Los Alamos National Laboratory," presentation to the panel, May 2011.

Shock Ignition

Shock ignition is yet another variant on the theme of slowing the main fuel implosion to minimize driver energy requirements, adding one more drive element to locally heat a limited quantity of fuel to thermonuclear burn conditions and then using alpha-particle deposition to propagate the burn wave into the assembled fuel mass. In shock ignition, rather than using a separate, high-intensity, ultrashort-pulse laser to heat the ignited volume, a short, high-intensity "spike" is added to the end of the main drive pulse shape to launch a very strong shock into the fuel. This inward-propagating shock collides with the outward-propagating shock constituted by the growing region of high-density fuel at the center, producing a spherical shell of fuel at a much higher temperature. The principle of shock ignition has been demonstrated in experiments on the OMEGA laser at LLE (Betti et al., 2007). Since the target has a smaller radius at the time that the high-intensity spike is required to launch the final shock, it is energetically advantageous if the laser optics can accommodate focal zooming or, alternatively, if the high-intensity spike can come from a separate set of lasers with smaller intrinsic spot size. An issue that arises with shock ignition is that the final, high-intensity spike exceeds the threshold for laser-plasma interactions, which can interfere with the desired effect (see further discussion in Chapter 4).

Z-Pinch Ignition

Z-pinch targets need to achieve the same overall fuel parameters—that is, sufficient temperature to initiate thermonuclear burn and area mass density to initiate alpha-particle bootstrap heating of the remaining fuel mass. Since the targets are typically cylindrical, the convergence is only two-dimensional and it is more difficult to meet the ρr criterion. Some target designs work on the hot-spot ignition principle, in which a small central mass is shock-heated to thermonuclear temperatures.

Alternatively, in magnetized-target fusion (MTF), the fuel mass is preheated by an energy source (e.g., a laser beam) to place it on a higher adiabat. Field coils are placed around the target to provide a seed magnetic field throughout the fuel volume. The magnetized, preheated fuel is then imploded at a lower implosion velocity than is used in hot-spot ignition to minimize driver energy requirements. The magnetic field is applied to inhibit fuel cooling during the slow implosion process (i.e., inhibit cross-field transport). The higher initial adiabat allows the magnetically insulated fuel to reach thermonuclear conditions at smaller convergence ratios. The principle of MTF has not yet been successfully demonstrated. MTF is normally considered more as an attempt to find an easier path to ignition rather than as a path to high yield and high gain, but recent numerical simulations indicate that high-gain MTF is possible using cylindrical implosions with a cryogenic DT layer (Slutz and Vesey, 2012).

What Determines the Degree of Fuel Burn and Gain?

Fusion yield Y scales strongly with capsule absorbed energy ($Y \sim E^{5/3}$), which implies there is a strong premium on efficiently delivering energy from the driver to the capsule. Energy must be absorbed symmetrically into the fuel to avoid instabilities. Each target design has different transport and deposition issues:

- Indirect drive (e.g., in the NIC at the NIF) requires transport of lasers through a background gas and delivery through laser entrance holes (LEHs) in the hohlraum (see Chapter 4). Most of the driver energy goes to heating the hohlraum wall and the dense plasma blown off the wall, so the process is inherently inefficient.
- Direct drive simplifies transport and focusing issues, but it is critical to avoid the generation of hot electrons (which cause fuel preheat) from laser-plasma interactions. This method is more efficient because it is direct, but symmetry and deposition physics are very important.
- Z-pinches require a direct electrical connection between driver and target through a recyclable transmission line (RTL). As the target implodes and the Z-pinch inductance increases, there may be potential loss regions. Because of the RTL, each shot requires the replacement of substantial structure.
- Heavy ions are charged particles that are susceptible to plasma instabilities when they are focused to the intensities required for ICF (>500 TW). Accelerators work best at low currents, so achieving a high power requires high particle energies, which makes their energy deposition range long. This complicates target design.

As noted above, fusion yield is calculated to scale as absorbed energy $E^{5/3}$, so delivering more energy to the target results in significantly higher yield. For the same driver energy, direct drive delivers more energy to the fuel than does indirect drive. Implicit in this yield-scaling is the fact that the increasing fusion energy output comes from burning more fuel. Burning more fuel requires compressing more fuel to near Fermi-degenerate conditions, which requires more energy to be absorbed by the target. Since most of the fuel mass is in DT at solid (ice) density, more fuel mass means targets of larger radius. Larger target radius has the additional benefit that it increases the inertial confinement time of the fuel mass (determined by the imploded fuel radius divided by the sound speed) and increases the burn-up fraction of the DT fuel disassembly. The burn-up fraction depends on the areal density of the fuel capsule:

$$f_b = \rho r/(\rho r + \beta(T))$$

where $\beta(T) = 5.5\text{-}6.5$ g/cm^2 for optimal burn conditions. For a burn-up fraction greater than about 1/3, ρr must be greater than about 3 g/cm^2.

All designs try to use driver energy efficiently; thus, they implode a cold mass of fuel isentropically and a small amount of fuel to high temperature—either by hot-spot ignition, fast ignition, or shock ignition. Instabilities can limit the propagation of burn from the ignition region to the remaining fuel. "Yield over clean" (YOC) is a measure of the deviation of experiments from ideal simulations.

Spectrum Output

The fusion reaction determines the initial partitioning of energy into alpha particles, X-rays, and neutrons. The spectrum of particles hitting the IFE target chamber wall is a function of the intervening materials, whether from the hohlraum, support structures (e.g., RTLs), or chamber fill gas.

Indirect-drive targets have high-Z materials in the hohlraum that emit copious X-ray radiation. Xenon gas can be used to absorb these X-rays and mitigate chamber wall damage (see Chapter 4). The xenon gas will get hot, but the hohlraum is believed capable of protecting the cryogenic fuel as it transits the chamber.

Direct drive usually assumes a vacuum in the target chamber, because the fuel pellet cannot be thermally insulated from a hot background gas. A shroud containing helium gas at low pressure and temperature has been considered, although it presents many difficulties. Even though the target is made of low-Z materials, there are still X-rays and ions that strike the wall and deposit their energy very locally. Magnetic diversion of ions is being considered in some designs to protect the chamber wall.

Z-pinch reactors would have yields above 1 GJ and RTL structures in the chamber.[5] This can lead to debris and shrapnel. The RTLs also can contain substantial residual magnetic field energy, which needs to be accounted for in determining which particles hit the wall. Thick, Li-containing liquid walls can be used to protect the chamber surface from short-range ions, neutrons, and X-rays.

Heavy-ion driver concepts are tending to use liquid walls and perhaps background gases. There do not appear to be any unique or particularly challenging aspects to the heavy-ion output spectrum as compared with laser direct-drive or indirect-drive systems.

---

[5] M. Cuneo et al., Sandia National Laboratories, "Pulsed Power IFE: Background, Phased R&D, and Roadmap," presentation to the NRC IFE committee on April 1, 2011.

Target Injection and Fabrication

For energy to be produced in a fusion reactor, the target (which is the fuel source) will be obliterated. Thus, for IFE to produce a steady flow of energy, a steady supply of new targets must be introduced into the system. The more frequently the targets are introduced and converted into energy, the more power is produced; and similarly, the more energy that is available in each target, the more power is produced. It is the details of these targets, and how efficiently the energy is released, that distinguish the different concepts for IFE. These differences and technical challenges are discussed in detail in Chapter 4.

How frequently targets can be introduced into the fusion reactor (the repetition rate) is determined by engineering practicalities of each fusion concept. The repetition rate for the concepts discussed here varies from 0.1 to 20 Hz. These values are calculated estimates; the technical challenges of delivering targets into the fusion chamber at these rates with the required precision, while preserving the integrity of the target, has been—in the absence of a comprehensive IFE program—only superficially addressed. Specific engineering concepts will require comprehensive testing to determine whether the proposed repetition rates, and subsequent power production, are feasible. Equally important is to understand whether any degradation to the configuration of the target during this injection process could reduce fusion performance below the calculated performance.

Operating a fusion reactor at a repetition rate of 20 Hz will consume 1.728 million targets per day. No credible process for cost-effectively producing this number of targets has been developed. Current ICF experiments show that there is a technical path for manufacturing targets that meet critical specifications; whether this technical path is a viable method for mass-producing targets remains to be established. These considerations are discussed next.

Target Injection

For laser-driven IFE, the target injection process poses four challenges: accuracy and repeatability (both spatially and temporally) of target placement, ability to track the target, target survival, and clearing of the chamber. These challenges are discussed in the following paragraphs.

A necessary condition for achieving the optimal energy output from each target is that the target be uniformly compressed by the laser beams. This requires the target to arrive at the same point in space and at the same instant as the multiple laser beams. For the direct-drive target, the target must be within 20 μm (rms between the centerline of laser beamlets to the centerline of the target). Concepts developed and tested as part of the High Average Power Laser (HAPL) program[6] (see Box 4-2 in Chapter 4) showed that a surrogate target could be repeatedly placed within 10 mm of target chamber center, where a final engagement system does the final pointing. For the indirect-drive targets currently under development, the target is required to be within 100 μm of the focus of the laser beam,[7] which appears to be within the capabilities of the system developed by the HAPL program; however, one difference between the direct- and indirect-drive approaches to fusion is that the indirect-drive approach has a higher gas pressure in the reactor chamber that may affect the repeatability of the injection process (Norimatsu et al., 2003). These are issues to be resolved in a technology development program.

The second challenge is the ability to track the target to make real-time, minor corrections to the pointing of the laser beams at the target. Here technical progress was achieved during the HAPL program by demonstrating the ability to track a target moving at 5 m/s and to steer beams in real time so as to engage it with ± 28 μm accuracy (Carlson et al., 2007). The system has been designed assuming an injection velocity of 50 m/s.

The third technical challenge is to preserve the target's critical specifications until the moment of the implosion. The problems are significantly different in this case for direct- and indirect-drive targets. For indirect-drive targets, the surrounding hohlraum will provide thermal protection. However, laser access to the target is through thin membranes (<0.1 μm thick) at each end of the hohlraum, and these holes will allow a sizeable heat load (both radiative and conductive) to be delivered to the target. The radiation portion of this heat load is reduced by the presence of internal shields within the hohlraum, which will also disrupt convective cells, but the conductive heat load is unaffected and the target's temperature is calculated to rise ~85 mK, which is less than the 100 mK ceiling specified in one system design.[8] The benefit of these structures to the target's preservation is appreciable; however, this benefit comes at the cost of a complex structure that needs to be built to high precision, and this precision must be maintained during the acceleration loads that the target experiences when it is injected into the reactor. These loads to the target assembly have been calculated and are stated to be acceptable.[9]

For direct-drive targets, target survival is the major challenge. The exact heat load to the target is strongly dependent on engineering parameters such as the gas pressure in the reactor chamber, the time the target is inside and exposed to the environment, and the temperature of the reactor; heat fluxes in excess of 1 W/cm^2

[6] J. Sethian, Naval Research Laboratory, "The HAPL Program to Develop the Science and Technologies for Direct-Drive Laser Fusion Energy," presentation to the panel on September 20, 2011.

[7] M. Dunne, LLNL, "LIFE Target System Performance," presentation to the panel on July 7, 2011.

[8] Ibid.

[9] Ibid.

to the target will compromise the target's performance (Tillack et al., 2010; Bobeica, Ph.D. thesis; Bobeica et al., 2005).

Multiple strategies are envisioned for minimizing the heat load; two possibilities are to add protective layers to the outer surface of the target and to minimize the gas pressure in the reactor (Petzoldt et al., 2002). Testing such strategies is a critical step in determining the engineering feasibility of the laser direct-drive fusion energy option.

Finally, it is necessary to clear the chamber of debris between shots. In the past, there has been a tendency to minimize this problem because the other issues appear so much more daunting. However, new concepts, higher repetition rates (with incrementally more mass injected into the chamber per unit time), and the possibility of increasing the gas pressure in the reactor to improve the durability of the reactor structure (high gas pressure will reduce the X-ray and ion-induced damage to the chamber wall) complicate the process of clearing the chamber.

Concepts for injecting targets for pulsed-power fusion energy are radically different and less fully developed than their laser-driven fusion energy counterparts. The signature difference is that targets are consumed at a rate of 0.1 Hz and that the target is a more massive structure (up to 50 kg) that includes transmission lines that couple the power to the target.[10] Removing spent targets and installing new targets will be done using automated machinery.[11] While this process is conceptually feasible, there remain substantial engineering considerations that need to be resolved to determine whether this process can be completed within 10 seconds.

The heavy-ion fusion energy concepts originated as a variation of laser-driven concepts in which the driver energy is supplied by heavy ions accelerated by a linear accelerator. Subsequently, a variety of target-design concepts have been proposed: an indirect-drive design (3-4 GeV Bi^{+1}); polar direct-drive design (3 GeV Hg^{+1}); and a single-sided direct-drive configuration (90 GeV U^{+4}).[12] The target-design concepts use indirect-drive, direct-drive, and single-sided direct-drive configurations. The target injection challenges are similar for heavy-ion and laser-driven fusion: the indirect-drive target benefits from the thermal shielding provided by the hohlraum, while the direct-drive target remains vulnerable to the hostile environment of the reactor chamber. Beyond these commonalities with laser-driven fusion, no target injection concept specific to heavy-ion fusion has been proposed.

[10] M. Herrmann, Sandia National Laboratories, "Z-pinch Target Physics," presentation to the panel on February 17, 2011.

[11] M. Cuneo et al., Sandia National Laboratories, "The Potential for a Z-pinch Fusion System for IFE," presentation to the panel on May 10, 2011.

[12] B.G. Logan, Lawrence Berkeley National Laboratory, "Heavy-Ion Target Design," presentation to the panel on July 7, 2011.

Target Fabrication

Before the targets can be injected into the reaction chamber they must be fabricated to tight tolerances, which requires a well-understood and reliable process that is suitable for mass production. The mass fabrication challenges posed for the different types of targets vary significantly, although there are technologies common to many of the targets that will benefit all concepts for fusion energy. In this section, the key challenges are outlined for the production of these targets for laser drivers, pulsed-power drivers, and heavy-ion drivers.

Targets proposed for each of the fusion energy concepts have equal mixtures of deuterium and tritium as the fuel. This fuel is confined in a spherical capsule for the laser-driven concepts and most of the heavy-ion concepts or in a conical "X-target" (see Figure 2-6) or cylindrical structure (see Figure 2-7) for direct-drive heavy-ion fusion and pulsed-power fusion, respectively. Fabrication of the conical and cylindrical structures appears to be straightforward, though the exact specifications are not yet well defined or tested. Fabrication of the spherical capsules is complicated—partially owing to the design and partially owing to the tight tolerances and stringent specifications. Researchers making these targets for the ICF and the HAPL programs produced targets with specifications that are acceptable for the laser-driven fusion concepts; however, it remains to be demonstrated that the fabrication process can be scaled to satisfy the requirements of an IFE program.

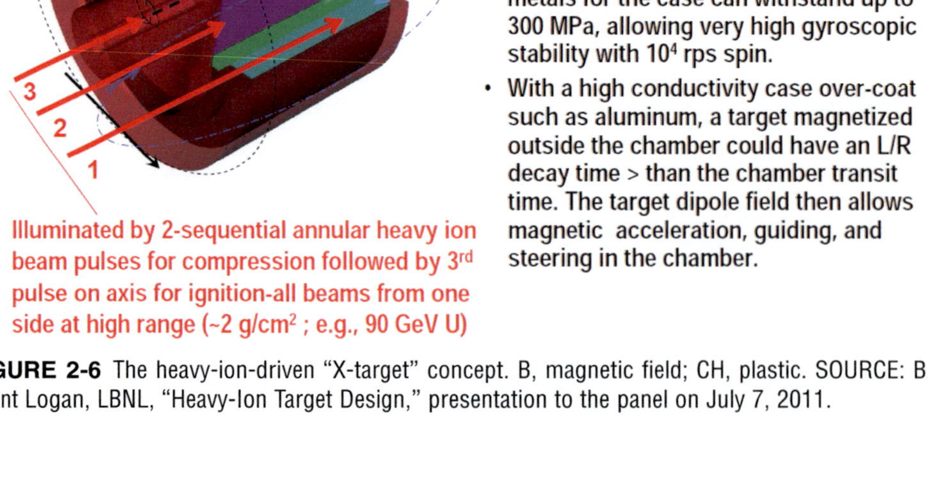

FIGURE 2-6 The heavy-ion-driven "X-target" concept. B, magnetic field; CH, plastic. SOURCE: B. Grant Logan, LBNL, "Heavy-Ion Target Design," presentation to the panel on July 7, 2011.

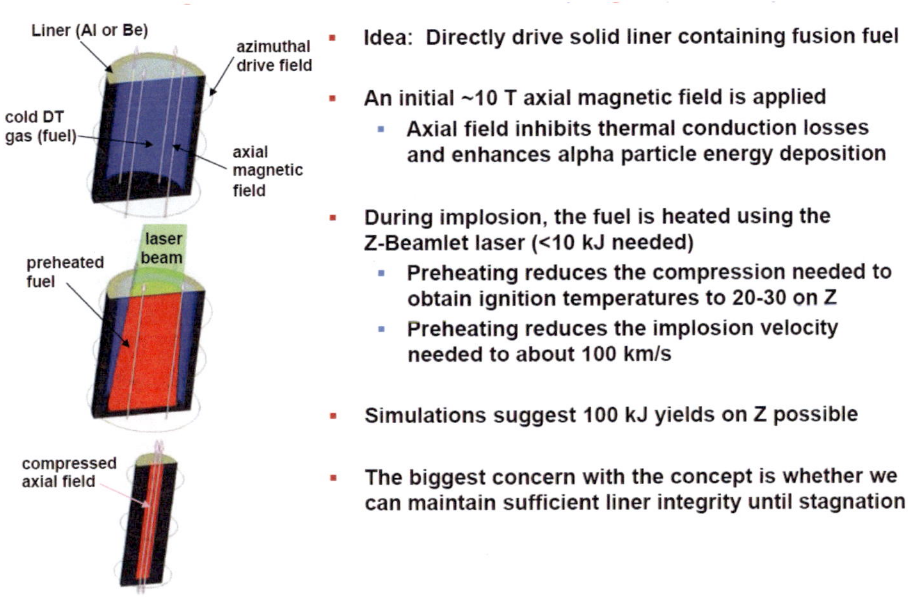

FIGURE 2-7 The cylindrical magnetized liner inertial fusion (MagLIF) target concept. SOURCE: S.A. Slutz, SNL, "Design and Simulation of Magnetized Liner Inertial Fusion Targets," presentation to the panel on May 10, 2011. Top image adapted from S.A. Slutz et al., Pulsed-power-driven cylindrical liner implosions of laser preheated fuel magnetized with an axial field, *Physics of Plasmas* 17:056303 (2010).

Indirect-Drive Targets

The indirect-drive targets proposed for laser-driven IFE (e.g., in the LIFE point design) are a modification of the target currently used at the NIF. The fundamental design is the same: DT fuel is contained inside a capsule that is supported inside a hohlraum. However, there are differences in both the capsule and the hohlraum. The capsule is a bilayered structure with an outer layer of high-density carbon (diamond) and an inner layer of low-density hydrocarbon foam. The hohlraum is elliptical (rather than cylindrical as is the NIF target) and made from lead rather than gold. Additionally, internal membranes ("shine shields") are introduced to prevent the capsule having a direct line of sight to the laser entrance holes in the hohlraum. The capsule is postulated to be manufacturable using a combination of microfluidic and vapor deposition techniques, and the DT fuel is added by drilling a hole 5 μ in diameter in the capsule and sealing it once the fuel is inserted. Cooling the target assembly liquifies the DT fuel, which is wicked into the foam layer to make a uniformly thick fuel layer. New technologies will be required to form the foam layer inside an existing capsule, and those technologies need to be consistent with a credible mass-production process.

Direct-Drive Targets

The direct-drive target proposed for fusion energy bears a close resemblance to the direct-drive target that is proposed for experiments at the NIF.[13] The fusion energy target is a spherical foam capsule that is slightly larger than the NIF direct-drive target. The outer surface of the foam capsule has a fully dense plastic overcoat (to retain the fuel) and a thin reflective metallic coating to reduce the radiative heat load to the ice. Additional outer layers may be needed to provide greater protection to the target when it is injected into the reactor chamber. The DT fuel is diffused into the plastic shell, and the target assembly is cooled to form the uniformly thick ice layer.

The manufacturing processes for both laser-driven target designs are scalable for mass production. However, it remains to be demonstrated that these processes can achieve the production yield required for a fusion plant given the specifications that are required. At this point, such processes are near[14] but have not yet been proven for mass production. Any changes in the target design to improve the implosion physics (resulting from experiments at the NIF) are likely to be dimensional changes that can be easily accommodated by the existing manufacturing process instead of changes in configuration that would require new technologies.

Two of the targets designs that are proposed for the heavy-ion driven fusion concept use indirect- and direct-drive implosion symmetries, so the manufacturing challenges are the same as for laser-driven fusion targets. A third more recently proposed target design is a single-sided direct-drive concept where liquid DT fills an X-shaped volume (two cones joined at the apex; see Figure 2-6). No production method has been proposed, nor are any tolerances proposed for the design, although it appears this target will have similar constraints and technical challenges as the other targets.

The pulsed-power fusion energy targets are distinctly different from the other fusion energy targets. There are multiple designs; one is a cylinder made from beryllium and filled with cryogenic DT gas. This target will be straightforward to manufacture and is considerably less complex than the other target designs. However, the additional components that are needed to inject this target into a

---

[13] P.B. Radha, University of Rochester, "Polar-Drive Target Design," presentation to the panel on July 7, 2011.

[14] J. Sethian, NRL,"The HAPL Program to Develop the Science and Technologies for Direct-Drive Laser Fusion Energy," presentation to the panel on September 20, 2011, and M. Dunne, LLNL, "LIFE Target System Performance," presentation to the panel on July 7, 2011.

pulsed-power fusion reactor must be better defined to fully evaluate the technological challenges to making the entire target assembly.[15]

Factors Most Likely to Determine the Cost of Targets

It is important to appreciate that the technologies for making most of the components of the targets exist already; targets are being successfully manufactured for the existing ICF program, and with a few exceptions, any changes to the target to adapt it for energy applications appear to be technically feasible.

Much of the cost of the ICF target today is due to the quality assurance process, in which each target must be thoroughly evaluated because the yield of acceptable targets is so low. Any future IFE technology program will need to evaluate whether current technologies can (1) produce a more consistent product and (2) maintain the high production yield when scaled to mass production.

The material and production costs for manufacturing the targets appear to be acceptable and will benefit from the economies of large-scale production if a viable process is developed. The costs for developing the manufacturing process and constructing the manufacturing facilities are less predictable, with the latter depending strongly on the former. However, these are one-time costs that when amortized over the number of targets that are produced during the projected lifetime of the plant will likely be a small component in the cost of each target.

A contributor to the cost of the target is the cost of the tritium fuel. Fusion energy has the appeal and requirement that tritium be bred in a reactor and be self-sustaining. Neutrons from the deuterium-tritium fusion process interact with a surrounding blanket of lithium/beryllium and produce proportional quantities of tritium. Once the plant is initially fueled with tritium, the cost of sustaining the fuel will be primarily the cost of extracting tritium from the by-products of the nuclear reaction and the cost of controlling the radiological hazards. (Deuterium, the other component of the fuel, is extracted from water.)

Tritium Inventory Considerations

A consideration for selecting a target production concept, and possibly even a fusion energy concept, is the amount of tritium that is required to maintain the power plant in constant operation. While tritium-breeding will allow a facility to be self-sustaining, the complexity of recovering tritium from the breeder and reactor-chamber effluent, and then refueling the targets, will scale with the complexity of the operation and amount of tritium in the facility.

---

[15] S.A. Slutz, SNL, "Design and Simulation of Magnetized Liner Inertial Fusion Targets," presentation to the panel on May 10, 2011.

Minimizing the amount of tritium in a power plant was an important consideration in designing the indirect- and drive-direct targets.[16] More ambitious ideas were proposed for the indirect-drive concept that will require additional scientific and technical development to realize: drilling a hole in the target to add the fuel (and then resealing the hole) and achieving a uniformly thick fuel layer by suspending the fuel as a liquid within a foam layer. Combined, they would reduce the tritium inventory to less than 1 kg[17] by recycling tritium through the facility in less than 8 hours. The first approach adds steps to the manufacturing process and should be technically feasible; the latter approach is also technically feasible, but it is unclear whether the liquid fuel can be cooled below its freezing point and still remain a liquid, which is what has to be done to achieve the gas density required in the capsule. If this is not possible, then an alternative and lengthier process is needed to form the ice layer, which would increase the tritium inventory.

Minimizing the tritium inventory was a less important consideration for developing the direct-drive target. In any case, target tritium inventory for the direct-drive targets is much higher than for the current indirect-drive configuration. About 10 times more tritium is present in this target than in the indirect-drive target. Additionally, tritium is diffused into the capsule instead of flowing through a hole, which takes 2 to 4 days because of the fragility of the target and the quantity of fuel that has to be added.[18] The process for forming the ice layer adds about 12 hours to the production cycle, which is the same process that the indirect-drive concept will use if it is not possible to subcool the liquid layer sufficiently to achieve the desired gas density.

Two main contributors to the total tritium inventory of an IFE plant will be these:

- The amount of tritium that is trapped inside the target during the target assembly phases and
- The amount that is entrained in the tritium-breeding and recovery processes (from the gaseous effluent from the reaction chamber).

At this stage, there is insufficient information to know the optimum balance between these sources and whether the effort to minimize the amount of tritium in the target assembly process is worth the added manufacturing and technical complexities.

---

[16] M. Dunne, LLNL, "LIFE Target System Performance," presentation to the panel on July 7, 2011.

[17] M. Dunne et al., LLNL, "Overview of the LIFE Power Plant," presentation to the panel on April 6, 2011.

[18] J. Sethian, Naval Research Laboratory, "The HAPL Program to Develop the Science and Technologies for Direct-Drive Laser Fusion Energy," presentation to the panel on September 20, 2011.

3

Proliferation Risks Associated with Inertial Fusion Energy and with Specific Target Designs

This chapter discusses the potential proliferation risks associated with inertial fusion energy (IFE). Many modern nuclear weapons rely on a fusion stage as well as a fission stage, and there has been discussion of the potential for nuclear proliferation—particularly vertical proliferation[1]—in a country where an IFE power plant is sited.

The panel begins by providing some background on nuclear proliferation and inertial confinement fusion (ICF) and continues with discussions of several related topics: classification concerns, the relative proliferation risk associated with different target designs, weapons production in ICF facilities, knowledge transfer, other proliferation risks associated with ICF, and, finally, the importance of international engagement on this issue.

CONTEXT AND HISTORICAL PERSPECTIVE

The term "nuclear proliferation" refers to the spread of nuclear weapons knowledge, technology, and materials to countries or organizations that did not previously have this capability. Proliferation has been of increasing concern in recent years, particularly following the successful detonation of a North Korean nuclear weapon, and the signals that Iran may also be pursuing an illicit nuclear weapons program. With the breakup of the Soviet Union, special nuclear material (SNM)

[1] Vertical proliferation refers to the enhancement of a country's capability to move from simple weapons to more sophisticated weapons.

became available at lightly guarded facilities; it is unclear how much was lost to theft, but proliferation concerns remain. Another concern arises from the many nuclear weapons in Pakistan, and whether they are controlled adequately.

Proliferation could occur in several ways: (1) the spread of knowledge about how to build nuclear weapons to other countries, (2) knowledge of—and access to—the physical technology used to construct nuclear weapons, (3) access to the materials from which a nuclear weapon could be constructed (e.g., SNM), and (4) access to people who have been engaged in nuclear weapons technology in other nations.

Because the first nuclear weapons were built using technology that was later adapted for use in civilian nuclear power plants and the civilian nuclear fuel cycle, the role that fission power could play in proliferation has been considered for decades. An international safeguards regime to detect attempts at proliferation is currently in place and operated by the International Atomic Energy Agency (IAEA). This regime, which is based on the Treaty on the Non-Proliferation of Nuclear Weapons (NPT), involves cooperation in developing nuclear energy while ensuring that nuclear power plants and fuel cycle facilities are used only for peaceful purposes.

The risk of nuclear proliferation could also be associated with ICF research facilities or, possibly in the future, IFE plants. For example, IFE plants and ICF research facilities provide an intense source of neutrons, which could, in principle, be used to generate ^{239}Pu from ^{238}U. In addition, information that could help countries develop more advanced boosted weapons or thermonuclear weapons could be gained from a thorough understanding of a fusion facility's operation.

While the effect of a fission-only weapon can be devastating, the development of two-stage (both fission and fusion) thermonuclear weapons can provide much higher yield per weapon. By using an ICF facility to improve its understanding of the physics of fusion, a nation might glean information useful in transitioning its weapons program into a much more complex, modern, and threatening system. In fact, the U.S. research program in laboratory-based ICF has been largely funded by the nuclear weapons program, because valuable information can be learned from ICF that can otherwise be learned only from nuclear testing.[2]

Because IFE is still at an early stage as a potential energy source, international treaties related to nuclear weapons and proliferation do not clearly apply to IFE at this time. However, given the value of ICF to the U.S. nuclear weapons program

[2] The moratorium on nuclear testing announced on October 2, 1992, by President George H.W. Bush and extended by the Clinton administration remains in effect. It was reinforced by the 1996 U.S. signing of the Comprehensive Nuclear Test Ban Treaty, which, however, has not been ratified by the U.S. Senate. The information gained by the nuclear weapons program is related to improving our understanding of weapons components built during the cold war, including the effects of aging on component performance.

and the programs of other nations, the applicability of some treaties to ICF has been considered.

The NPT does allow for laser fusion experiments, both in states that already have nuclear weapons and those that do not. As noted in 1998, this position is based on the unopposed, U.S. unilateral statement at the 1975 NPT Review Conference stating that "nuclear reactions initiated in millimeter-sized pellets of fissionable and or fusionable material by lasers or by energetic beams of particles, in which energy releases, while extremely rapid . . . are nondestructively contained within a suitable vessel . . . [do] not constitute a nuclear explosive device within the meaning of the NPT . . ." (U.S. DOE, 1995). Even so, the status of pulsed-power fusion experiments under the NPT remains unclear (Paine and Mckinzie, 1998).

In the 1990s, there was discussion in the United States about whether the Comprehensive Nuclear Test Ban Treaty (CTBT) also banned the use of ICF.[3] Ultimately, the Clinton administration took the position that ICF is not a prohibited activity under the CTBT (Jones and von Hippel, 1998), and this position continues to be that of the Obama administration. However, some experts still debate the applicability of this treaty to ICF (Paine and McKinzie, 1998).

ICF research has received a great deal of specifically directed funding in the United States in recent years, even though IFE per se has not. This research is funded primarily through the U.S. nuclear weapons program, which envisions using ICF experiments and modeling as a method of verifying codes and calculations related to the current U.S. nuclear weapons stockpile. Because many of the topics involved in ICF are related in some way to nuclear weapons, much of the work is classified. The next section provides a brief introduction to the history and current status of the classification and declassification of various ICF concepts.

CLASSIFICATION: ICF AND IFE

The primary reason stated by the U.S. government for classifying information related to ICF is to protect information relevant to the design of thermonuclear weapons. The possibility of using lasers to ignite fuel was first considered by the Atomic Energy Commission (AEC) and the national weapons laboratories in the early 1960s. At that time, concerns about the potential for laser fusion weapons as well as close ties between ICF concepts and nuclear weapons design (particularly physics and simulation codes) led the AEC to classify research on ICF. The first classification guidance for inertial confinement fusion information was issued in 1964. Initially, all aspects of ICF were considered to be classified.

[3] It should be noted that the United States is not currently a party to the CTBT but as a signatory is bound not to act in violation of the fundamental restrictions of the CTBT.

Declassification of fusion concepts began slowly in the 1970s, and by August 1974, essentially all work with directly irradiated fusion targets was declassified. After a long pause, declassification began again in the late 1980s and continued through the early 1990s. Most notably, in late 1990, an Inertial Confinement Fusion Classification Review was requested by the Secretary of Energy with the intent of eliminating unnecessary restrictions on information relevant to the energy applications of inertial confinement fusion. The panel included representatives from the Department of Energy (DOE) national laboratories, the Department of State, the Arms Control and Disarmament Office, and other stakeholders, and the report was issued on March 19, 1991. The key panel recommendations included these: (1) "For laboratory capsules absorbing <10 MJ of energy and with maximum dimension <1 cm, all information should be declassified with some exemptions" and (2) "Some Centurion-Halite declassification would be desirable to gain the scientific credibility needed to advance the energy mission of ICF" (U.S. DOE, 2001). Later, on December 7, 1993, nearly all information on laboratory ICF experiments was declassified.[4]

At present, much of the information related to ICF targets has been declassified, with several notable exceptions. First, some aspects of computer codes and certain target designs remain classified, as well as the details of some historical experiments related to ICF (in particular, the Centurion-Halite program). Some aspects of classified targets are discussed in the classified Appendix F.

Whether or not aspects of ICF are classified is highly relevant to the future of IFE. If essential parts of an IFE plant are classified, this could create significant complexities for commercialization. Although some commercial facilities rely on classified concepts (such as those involved in the enrichment or reprocessing of nuclear fuel), there are likely to be export controls or specific regulations involved in dealing with this situation.

It is important to realize that classification or export controls could themselves indirectly cause proliferation risks if denial of information, technology, or materials causes some nations to mount covert programs or withdraw from the NPT.

There are four possible scenarios for future classification of IFE concepts. The first possibility is simple—the target will be classified or other key aspects of the concept will be classified. The second possibility is that the target is unclassified, but the expertise needed to make or assess it will involve classified information or codes. A third possibility is that other parts of the plant (e.g., lasers) will be considered to be dual use and subject to export controls. Any of these three outcomes could be very troublesome at a commercial plant. On the other hand, a fourth possibility is that the target and expertise will be unclassified, and none of the key elements of the plant are subject to export controls. If this is feasible, then

[4] R. Johnson, LLNL, "The History of ICF Classification," a document provided to the panel on February 24, 2011.

it would be the simplest configuration and a highly desirable goal for the future commercialization of IFE.

PROLIFERATION CONCERNS ASSOCIATED WITH DIFFERENT IFE TARGET CONCEPTS

Any kind of ICF seeks to achieve thermonuclear ignition and burn. As noted previously, this goal relates ICF to thermonuclear weapons, and for this reason ICF (whether in a research facility or a power plant) is seen to pose some proliferation risk. However, this risk is mitigated by the fact that (1) nuclear weapons are much larger than ICF targets and (2) their operation presents some different engineering challenges.

Indirect-drive targets are associated with some proliferation concerns because the physics involved is more closely related to the physics associated with thermonuclear weapons than is the case with direct drive. In particular, the functioning of indirect-drive targets involves the use of X-rays in the hohlraum to drive the capsule implosion. ICF using indirect drive was declassified in 1991.

In any case, the processes involved in heavy-ion deposition (for heavy-ion-driven fusion) and the beam-plasma interactions that occur in direct-drive capsules are physically much more remote from conditions in existing thermonuclear weapons. In addition, these processes do not relate to any feasible design for a weapon that the panel is aware of. For these reasons, it is the judgment of the panel that heavy-ion fusion and direct-drive fusion pose (arguably) fewer proliferation concerns.

The Z-pinch fusion concept is likewise remote from existing weapons. However, during the cold war, the Soviet program in explosively driven magnetic implosion (MAGO) progressed further than any other approach to pure fusion, though like all such approaches, it was still very far from ignition (Garanin et al., 2006; Velikhov, 2008). Since the 1990s, Los Alamos National Laboratory and the All Russian Research Institute of Experimental Physics (VNIIEF) have carried out joint experiments on MAGO (Lindemuth et al., 1995).

In the future, as processing power for desktop and academic computers continues to increase, and as knowledge of plasma physics continues to accumulate in the open literature, many of these concerns may become less relevant, including the proliferation risk distinction between indirect drive and other forms of ICF that might be used for IFE. Enough physics knowledge may accumulate in the public arena that the use of indirect-drive IFE would not be able to add much to publicly available knowledge. In such a world, codes would be classified according to their direct use for (and calibration from) nuclear weapons, not according to the physics that they model. However, if an IFE plant were to rely on classified codes for target design or other operational aspects, and knowledge of these technologies could be used to gain information about the codes' details, proliferation would be a concern.

CONCLUSION 3-1: At present, there are more proliferation concerns associated with indirect-drive targets than with direct-drive targets. However, the spread of technology around the world may eventually render these concerns moot. Remaining concerns are likely to focus on the use of classified codes for target design.

WEAPONS MATERIAL PRODUCTION AT IFE PLANTS

One of the key proliferation risks associated with any fusion plant (ICF or magnetic confinement fusion) is that it is possible to use the plant to create materials that are essential for the construction of nuclear weapons. These materials fall into two primary categories: special nuclear materials and tritium. Both types of material can be produced without the use of fusion facilities, but commercial fusion plants may be a more convenient source for these materials for those who cannot acquire them easily in another way. The potential for the production of each type of material is discussed next.

Special Nuclear Materials

As noted previously, it is technically possible to utilize the significant neutron flux emanating from a fusion reactor core to produce ^{239}Pu from ^{238}U. To accomplish this task covertly, it would be necessary to:

- Move quantities of uranium into the immediate vicinity of the fusion core and
- Acquire technology for—and construct—the appropriate reprocessing facilities to separate the plutonium from the uranium and fission products.

The first task is likely to be operationally cumbersome. In addition, the transfer of large quantities of uranium into and out of a fusion power plant would likely be detectable, because such conveyance would not be a normal operation for such a plant. The development and construction of a reprocessing facility—assuming that it had not already been built and brought into operation—would also be necessary. The technology is not new, but it requires significant radiation-handling capability. The construction and operation of such a facility would probably be detectable by the current safeguards regime.

Overall, the panel judges that the construction and diversion of an IFE plant in this fashion is not the simplest path for a host state to produce SNM. Research reactors and commercial nuclear plants capable of serving the same purpose (irradiation of uranium for plutonium production) exist in many nations. However, a previously built and operating fusion plant could serve as a path of opportunity for a nation interested in developing weapons. Such facilities may therefore have to be subject to inspection to assure that they would not be so used, and to IAEA safeguards in states that do not already have nuclear weapons.

However, if terrorists were to seize an IFE plant, it could provide them with neutrons for the production of material to make a weapon of mass destruction. In this case, any facility capable of producing neutrons could be useful, but it is possible that no better solution would be available. Nonetheless, as noted above, an effective form of reprocessing would still be needed to isolate the plutonium.

For these reasons, the panel believes that a fusion plant raises fewer proliferation concerns than a fission plant with respect to the production of nuclear materials. However, in a region free of nuclear facilities, siting of a fusion plant could increase the proliferation risk in that region if the fusion plant were totally exempt from inspection by the IAEA or other international body. A hybrid fusion-fission plant would have the proliferation disadvantages and the economic problems of both technologies.

Tritium

In order to fuel itself, a functioning IFE plant would likely be designed to continually breed a stream of tritium in vast amounts: about 60 kg per year for a plant of 1 GW (thermal) capacity. Tritium not only is an essential fuel for a fusion power plant, but it also can be used in part to fuel modern, boosted fission weapons or thermonuclear weapons.

The diversion of some portion of the substantial tritium stream would be relatively straightforward, but such diversion does not necessarily pose a significant proliferation threat per se. However, for a state already possessing nuclear weapons the diversion of only a few grams of tritium would be significant and would be difficult to detect. In addition, tritium can be produced in other ways if a state needs it. To date, tritium for nuclear weapons and other purposes has been produced using fission reactors.

With current technologies tritium alone, unlike SNM, cannot be used to build a nuclear weapon, and only a host state with relatively advanced capabilities would find such a stream of tritium to be useful. Indeed, for primitive nuclear weapons, tritium does not need to be used at all. However, if a significant diversion of tritium is observed, then it could be a signal to the international community that the host state is considering increasing its nuclear capability to include more advanced weapons using boosting or thermonuclear burn.

KNOWLEDGE TRANSFER AT ICF FACILITIES

A second path for a potential proliferator might be the covert acquisition of key information about fusion, drawing on knowledge gained from operating a fusion facility. This path is discussed separately for research facilities and energy facilities in the following sections.

Inertial Confinement Fusion Research Facilities

Research facilities—such as the National Ignition Facility (NIF)—pose different proliferation concerns than a fully functioning inertial fusion power plant, and the concerns associated with a host country misusing a research facility are likely to be greater than those associated with a fusion power plant. A fusion research facility is designed for the purpose of increasing physics understanding on a range of topics, not for a specific function (i.e., energy production). A power plant, however, is likely to be highly specialized and not designed with the flexibility inherent in a research machine. In addition, research facility diagnostics by their nature will provide hints about the underlying physics that power plant diagnostics may not.

If considered fully, the proliferation risk associated with a research facility can go beyond the physical presence of the facility in one nation or another. Research facilities may cater to a range of scientific interests beyond the needs of either the power generation community or the weapons community. For example, the NIF provides the plasma physics community with a highly effective experimental test and validation for a number of codes and theories that may indirectly or directly relate to the physics required for an understanding of thermonuclear weapons. Because the research community is intrinsically both open and international, such an improved understanding of plasma physics could provide a range of potentially useful information to a proliferator.

This increase in understanding is unlikely to stop, regardless of U.S. decisions. In the coming decades, both experiments and simulation in research facilities worldwide are likely to surpass current U.S. capabilities. For example, continuing increases in computing speed and understanding in the open research community could result in extremely capable physics codes.

However, it should be clear that information about physics is not the same as information about weapons design. For a nation that has never successfully (or unsuccessfully) detonated a thermonuclear weapon, no fusion research facility or power plant can adequately replace experimental physics and engineering knowledge gained from nuclear testing.

IFE Power Plants

An IFE power plant, as noted above, is unlikely to be highly flexible, and a research facility is likely to provide more information to a potential proliferator. By the time a design is commercialized, the physics will likely have been well understood (or engineered around), and the designs of the individual components will have been optimized to the extent possible for power production. In addition, the diagnostics will be likely to be optimized for the needs of a power plant operator, not for the needs of a physicist attempting to learn useful weapons information.

However, knowledge transfer remains a concern if an IFE power plant is deployed overseas in a country where proliferation is a concern, because local expertise will be needed to operate the plant. The plant may not yield useful information about the physics involved in the reaction, but could provide information about energies needed and other technological details that must be known to obtain ignition in a fuel pellet. Moreover, personnel would gain practical experience in handling tritium. Whether this knowledge would be greater than that obtainable in the open literature is unclear.

CONCLUSION 3-2: The nuclear weapons proliferation risks associated with fusion power plants are real but are likely to be controllable. These risks fall into three categories:

- **Knowledge transfer,**
- **SNM production, and**
- **Tritium diversion.**

CONCLUSION 3-3: Research facilities are likely to be a greater proliferation concern than power plants. A working power plant is less flexible than a research facility, and it is likely to be more difficult to explore a range of physics problems with a power plant. However, domestic research facilities, which may have a mix of defense and scientific missions, are more complicated to put under international safeguards than commercial power plants. Furthermore, the issue of proliferation from research facilities will have to be dealt with long before proliferation from potential power plants becomes a concern.

ICF FOR OTHER PURPOSES

One proliferation concern associated with ICF is the potential for the development of a laser fusion weapon, as discussed briefly in the section on classification earlier in this chapter. However, owing to the size, complexity, and energy requirements of existing or planned driver systems, the panel does not consider this to be a credible and immediate concern with respect to current concepts for inertial fusion energy, such as laser-driven fusion energy. However, in the distant future, advances in laser technology could change this picture.

In a 1998 declassification decision, DOE stated that "the U.S. does not have and is not developing a pure fusion weapon and no credible design for a pure fusion weapon resulted from the DOE investment" (U.S. DOE, 2001). According to information released after the cold war, the Soviet experience was similar. However, this concern might someday materialize with currently unforeseen technology developments. For this reason and to alleviate any current concerns, it will be important to address the possibility (or impossibility) of pure fusion weapons in policy discussions and in the safeguards regime.

THE IMPORTANCE OF INTERNATIONAL ENGAGEMENT

As described in the previous sections, there are proliferation risks associated with the use of ICF facilities around the world, and—should IFE concepts prove to be fruitful—with IFE plants themselves.

Managing proliferation, whether it is associated with fission concepts or fusion concepts, is intrinsically an international problem. While one country may not allow the export of certain technologies, other countries that do not consider the technology as sensitive may choose to allow it. In addition, the result of proliferation—the successful construction of a nuclear weapon by one more state—is international in its consequences.

For this reason, preventing proliferation associated with fusion energy requires international agreement on methods for managing the risks of the technologies involved, including safeguards. The IAEA defines the purpose of its safeguards system as follows:

> to provide credible assurance to the international community that nuclear material and other specified items are not diverted from peaceful nuclear uses. Towards this end, the safeguards system consists of several, interrelated elements: (i) the Agency's statutory authority to establish and administer safeguards; (ii) the rights and obligations assumed in safeguards agreements and additional protocols; and (iii) the technical measures implemented pursuant to those agreements. These, taken together, enable the Agency to independently verify the declarations made by States about their nuclear material and activities.

This safeguards system has been in place for decades to verify compliance with the NPT for fission plants and fuel cycle facilities around the world. If new facilities that also pose a proliferation risk—such as fusion facilities—were to be deployed around the world, it would be sensible to either include them in the current regime or to design a similar safeguards regime for them.

Of course, these safeguards would need to take into account the design of a particular fusion power plant. Although numerous design concepts have been advanced,[5] the panel did not see any credible, complete power plant designs. This

[5] See, for example, "OSIRIS and SOMBRERO Inertial Fusion Power Plant Designs," DOE/ER-54100-1, March 1992, and "Inertial Fusion Energy Reactor Design Studies Prometheus-L and Prometheus-H," DOE/ER-54101, March 1992.

has benefits, because it provides an opportunity to consider "safeguardability" directly in the initial design of a fusion power plant.

Early international discussions on this topic could be very helpful in reaching an international consensus on the key proliferation concerns associated with the use of inertial fusion power plants as well as how to manage these concerns (Goldston and Glaser, 2011).

CONCLUSION 3-4: It will be important to consider international engagement regarding the potential for proliferation associated with IFE power plants.

ADVANTAGES AND DISADVANTAGES OF FUSION PLANTS WITH RESPECT TO PROLIFERATION

Proliferation is most tied to access to SNM, e.g., using enrichment processes. Richard Meserve[6] recently wrote, "There is no proliferation risk from the [fission] reactors. Proliferation risks can arise from enrichment facilities because the technology could be used for weapons purposes" (Meserve, 2011). An advantage of fusion plants with respect to nonproliferation is that SNM will not be used in the plants and SNM will not be accessible from the waste products, as it is from fission plants. This lack of direct access to SNM is the major nonproliferation advantage of a fusion plant.

The disadvantage of inertial fusion power plants is that they allow access to knowledge and experience with fusion, which will necessarily increase with the design and operation of such plants. The latest nuclear weapons use fusion as a major source of the explosion energy. These concerns were outlined in one presentation by an official (Massard, 2010):

> As an EU [European Union] requirement, we keep a clear separation between IFE and "sensitive" weapons science (nonproliferation)
>
> - No use of weapons codes in the European programs
> - No benchmarking of physics code with weapons code
> - Not in favor of indirect drive capsule option in the European program for sensitivity issues

European countries have strong collaborations in ICF (e.g., HiPER). The French are building a laser fusion facility, LMJ, which is broadly similar to the NIF and which will be the most capable driver available in Europe. As a matter of policy, these programs will pursue indirect-drive ICF but do not intend to pursue indirect drive for IFE (Massard, 2010) because of the perceived proliferation risk. The

[6] Former Chair of the U.S. Nuclear Regulatory Commission and chair of the IAEA safety advisory group.

United Kingdom participates in LMJ and HiPER and also actively participates at the NIF in the United States, and in the latter context is pursuing indirect-drive ICF.[7]

The Russian program in pure fusion evolved historically from the pre-1991 Soviet nuclear weapons program (Velikhov, 2008). Its major emphasis is on magnetic confinement fusion, which is not within the scope of this report. In ICF, two methods have received continuing attention in Russia: laser fusion and magnetized target fusion (MTF). Although research supporting ICF development is ongoing with smaller lasers (Kirillov et al., 2000; Belkov et al., 2010), Russia currently has no laser facility comparable to the NIF or LMJ[8] and is unlikely to achieve laser-driven ignition in the near future. As for magnetized target fusion, the Russian MAGO concept has been widely advertised, and, as mentioned, joint work with LANL is ongoing. The proliferation risks of the MAGO MTF concept have been discussed in detail (Jones and von Hippel, 1998). Little concern about the potential for proliferation in MAGO is evident in Russian publications and policy. Indeed, in general, different countries have different classification policies.

[7] J. Collier, UK Science and Technology Facilities Council, "Recent Activities and Plans in the EU and UK on Inertial Fusion Energy," briefing to the NRC IFE Committee, June 15, 2011.

[8] A news report in August 2011 suggests that plans for an NIF-class laser at VNIEFF are once again going forward, with commissioning expected in 2017; however, the stated purpose is stockpile stewardship, not ICF. See http://english.ruvr.ru/2011/09/30/57370758.html.

4

Evaluation of ICF Targets

LASER-DRIVEN, INDIRECT-DRIVE TARGETS

Current Status

No laser fusion target has yet achieved ignition or breakeven,[1] but current understanding leaves open the possibility that given time, funding, and the existence of alternative design options with sufficient margin for ignition and a gain of one, ignition might eventually be achieved.

The current U.S. program aimed at achieving ignition, the National Ignition Campaign (NIC), lays out a path via laser indirect drive (ID), and significant progress has been made along that path, although not enough either to demonstrate success or to conclude that ignition cannot be achieved. It is the understanding of this panel that the current program plan anticipates a demonstration of ignition sometime after the beginning of FY2013, although the planning document scheduled that event for the end of FY2012. The closest Level 1 milestone as of this writing is to achieve, in FY2012, significant alpha-heating of a capsule's fuel. The expected signature of such an event is the production of at least 10^{16} deuterium-tritium (DT)-equivalent neutrons. The significance of this milestone is that it would indicate that fusion bootstrapping of the ion temperature in the capsule fuel had occurred—a prerequisite to achieving fusion ignition and energy gain.

[1] Breakeven occurs when fusion gain equals unity—that is, when the fusion energy released in a single explosion equals the energy applied to the target.

The NIC Rev 5.0 target is designed to operate using indirect drive of a frequency-tripled (3ω) laser to reduce the negative effects of laser-plasma interactions (LPI) (see Box 4-1).

Box 4-1
Laser-Plasma Interactions

In laser-driven inertial confinement fusion (ICF), the capsule implosion is driven by thermal pressure.[1] Thus, the incident laser energy must be absorbed by matter and thermalized, either in the outer shell of the capsule (direct drive) or in the inner walls of the hohlraum (indirect drive), which become plasmas. The variety of LPI that take place when an intense laser pulse hits matter have been studied for more than 50 years; they have been a key limiting factor in laser ICF, and are still incompletely understood.

LPI that absorb and thermalize laser energy are desired. Undesirable, parasitic LPI include backscattering of laser light, which can result in loss of energy; cross-beam energy transfer among intersecting laser beams, which can lose energy or affect symmetry; acceleration of suprathermal "hot electrons," which then can penetrate and preheat the capsule's interior and limit later implosion; and filamentation, a self-focusing instability that can exacerbate other LPI.

LPI are worse at longer laser wavelengths, so all modern drivers currently operate in the "blue" (3ω Nb:YAG at 353 nm) or ultraviolet (KrF at 248 nm). Moreover, lasers can be modulated so as to substantially ameliorate parasitic LPI by spectral broadening, spatially incoherent filtering, and/or polarization diversity, and great progress has been made over several decades on all the main kinds of laser drivers on such beam smoothing.[2] Since LPI are threshold effects, target designers attempt to keep laser intensities below the threshold of major harm. However, neither fundamental understanding nor simulation are good enough to do so a priori; well-diagnosed experiments remain essential for LPI control.[3]

LPI are currently important in the National Ignition Campaign (NIC) indirect-drive targets. Overall, backscattered light losses appear to be 10-15 percent of the incoming laser energy; however, the inner beams backscatter more because of their greater path length in the hohlraum plasma. Stimulated Raman scattering (SRS) of the inner beams appears to play a significant role in causing drive asymmetry and hohlraum temperature deficits.[4] The asymmetry has been controlled by the use of cross-beam energy transfer mediated by Brillouin scattering, but fundamental understanding and simulation of this effect are incomplete, and its repeatability has not been established experimentally. Experiments so far are said to indicate that hot electrons are below the design threshold, but more diagnostics are needed, because hot electrons, if actually present, could explain the currently observed anomaly in capsule adiabat. Furthermore, other laser-produced sources of preheat, such as gold M-band emission, will require quantification in this new cross-beam environment.

Rapidly increasing computer performance has enabled LPI calculations that were unimaginable just 12 years ago, but full-scale National Ignition Facility (NIF) simulations remain beyond reach.[5] The Lawrence Livermore National Laboratory (LLNL) typically performs single- or multiquad simulations using pF3D on the largest advanced simulation and computing (ASC) platforms. Improvements in hohlraum modeling have changed plasma conditions and the location of backscatter in LPI simulations, bringing them into better agreement with measurements. Recent simulations show that overlapping quads and spatial nonuniformities act to increase laser reflectivity. Simulations have suggested potential ways to mitigate the effect of overlap beam intensity on SRS, including changing the hohlraum aspect ratio and changing the pointing of inner cone quads. Substantial computational and experimental resources are being devoted to LPI issues within the NIC.

LPI for direct-drive targets are under experimental and theoretical study at the Laboratory for Laser Energetics (University of Rochester) (LLE);[6] the most important effect appears to be cross-beam energy transfer, which results in 20 percent energy losses in capsule experiments on OMEGA. The relatively short beam paths in coronal plasma suggest that other LPI, and hot electrons, may be controllable in the extrapolation to ignition targets for direct drive, though most of the key experiments remain to be done. However, the greater laser intensities needed for shock ignition may cause harmful LPI; this must be studied. OMEGA EP[7] will be an important platform for studying direct-drive LPI issues at inertial fusion energy (IFE)-relevant plasma scale lengths. Naval Research Laboratory (NRL) is performing complementary LPI experiments at 248 nm on Nike.[8] Two-plasmon decay experimental data seem to agree with thresholds calculated using simple plane-wave-based threshold formulas, confirming the classical wavelength scaling. In direct drive, the initial target aspect ratio can be modified to limit the intensity and mitigate LPI risk at the penalty of greater sensitivity to Rayleigh-Taylor hydroinstabilities.

Increased LPI intensity thresholds and greater hydrodynamic efficiency for short wavelengths should combine to give better overall stability in direct-drive implosions. The NRL baseline shock ignition target is above the two-plasmon decay threshold during compression (Liu and Rosenbluth, 1976). Extending the Nike laser to 20 kJ would provide a useful capability to study LPI and hydrodynamics at 248 nm in IFE-relevant scale-length plasmas and compare them with OMEGA extended performance and NIF data.

Plasma physics, including LPI, involves many degrees of freedom on a huge range of length scales; moreover, nonlocal propagation by electromagnetic fields and fast electrons are important. For these reasons, a priori simulation of a full-scale target will be impossible for the foreseeable future, although impressive simulations are now feasible for fundamental processes and small-scale regions. Future development of subgrid and mesoscale modeling on full-scale systems would help to understand the experiments and support better target design, but would require a large effort to create and perfect.

---

[1] Radiation pressure of the laser light itself is too small by many orders of magnitude.

[2] D. Montgomery, LANL, "Overview of Laser Plasma Instability Physics and LANL Understanding," presentation to the panel on September 21, 2011.

[3] M. Rosen, LLNL, "Understanding of LPI and Its Impact on Indirect Drive," presentation to the panel on September 21, 2011.

[4] Ibid.

[5] D. Hinkel, LLNL, "State of the Art for LPI Simulation," presentation to the panel on September 21, 2011.

[6] D. Froula, LLE, "Laser-Plasma Interactions in Direct-Drive Implosions," presentation to the panel on September 21, 2011.

[7] OMEGA EP (extended performance) is an addition to OMEGA and extends the performance and capabilities of the OMEGA laser system. It provides pulses having multikilojoule energies, picosecond pulse widths, petawatt powers, and ultrahigh intensities exceeding 10^{20} W/cm^2.

[8] A. Schmitt, NRL, "Assessment of Understanding of LPI for Direct-Drive (KrF)," presentation to the panel on September 21, 2011.

Recent and Upcoming Work

Recent work on indirect-drive laser fusion has brought the NIC program to the point where it has transitioned from preparation for the actual ignition campaign to the campaign itself. The latter involves optimization of a set of parameterized characteristics of the target and laser system in order to achieve conditions under which ignition could be anticipated to occur; the development of these "tuning parameters" has itself been one of the areas of development, in part because most of the tuning campaigns will require the use of specially designed capsules to enable data acquisition of the type and accuracy needed for that specific campaign.

Four key input variables are to be optimized in the NIC tuning campaigns:

- The implosion adiabat (usually designated α), which strongly affects the resistance of the capsule to implosion;
- The implosion velocity *V*;
- The amount of capsule material involved in mixing across the single interface characteristic of this class of capsule designs, *M*; and
- The overall shape of the implosion, which is characterized by a dimensionless parameter *S*.

These tuning campaigns are expected to use what are termed "keyhole" targets, backlit gas capsules, "symcap" capsules, and reemission capsules. Ignition is neither expected nor desired in these types of capsules, although tritium-hydrogen-deuterium (THD) capsules, which are intended for use in many of the preignition integrated experiments, utilize the ignition design but incorporate less DT thermonuclear fuel in favor of the less reactive HD. The use of THD capsules is expected to allow collection of data with which to confirm or calibrate calculations of the nuclear performance of the optimized implosion system (laser pulse + hohlraum + capsule design). Calibration of the nuclear diagnostics is planned using capsules of the so-called "exploding pusher" design.

The work mentioned thus far has all been accomplished at the NIF facility at LLNL. Additional preparations for optimization and testing of ignition capsules have been carried out at other laser facilities, notably the OMEGA laser at the University of Rochester's (LLE). One aspect of this work has investigated some of the problematic aspects of LPI. Experiments at LLE have also facilitated the development and porting of diagnostics to the NIF and have provided data on the operation of noncylindrical "rugby" hohlraums;[2] experiments are

[2] Rugby hohlraums are shaped not like a cylinder but like a rugby ball, with a wall having a tapered curve.

planned to provide similar data on the efficacy of "P2" laser entrance hole (LEH) shields.[3]

If ignition can be achieved on the NIF, target simulations presented to the panel suggest that optimization of the tuning parameters and increases in the driver energy could result in gains of between 50 and 100 at some future facility.

Evaluation and Discussion of Remaining R&D Challenges

It is too early in the experimental campaign to evaluate the performance of the NIC ignition target design. However, information already in hand does indicate some potential problem areas, which could become showstoppers. They are discussed individually below.

Implosion Velocity

Perhaps the most critical discrepancy is that the measured implosion velocity of nonoptimized capsules is ~10 percent lower than the calculated velocity, even early in the implosion. The fact that related quantities, such as capsule bang time, are likewise delayed compared to expectations confirms the interpretation of the velocity measurements. Possible explanations offered at the time the panel received its briefings are that the calibration of the hohlraum temperature measurement (Dante X-ray flux diagnostic) was incorrect, or that the opacity of the Ge dopant in the capsule wall (to reduce early-time heating of the interior portions of the capsule) was higher than expected. Plans are in place to explore these hypotheses by checking the calibration in question and testing capsules without that dopant for comparison.

The principal means available to increase the implosion velocity is to increase the laser drive energy. Greater drive energy would, however, also increase the preheating from LPI, which, as discussed below, does not appear to be well understood. A path forward is thus not guaranteed.

Implosion Symmetry

The panel was told that there are some concerns about early-time imprinting of drive asymmetries based on observations of reemission targets. Furthermore, the overall implosion symmetry of baseline targets was routinely more prolate

[3] P2 refers to the type of departure from sphericity that the shields are intended to reduce. A nearly spherical shape with azimuthal symmetry is often represented mathematically using Legendre polynomials, and P2 is the standard means of referring to the second Legendre polynomial, which is needed to describe a shape that has been described as a "sausage."

than predicted. Acceptable symmetry was obtained using interbeam energy transfer between outer and inner laser cones, but at present this process has not been successfully incorporated into the design simulations used to predict target performance. The consensus of the panel is that this situation may be a further indication of unknown LPI processes in the hohlraum or of other predictive inadequacies.

Mix

The prediction of mix across shocked interfaces and during convergent implosions has been a very active and controversial area of research in many technical communities for many years. Approximate simulations of mix are possible and are routinely included in some target simulations, but the calculated mix—and therefore its calculated effects—is recognized to be unreliable. Moreover, data to validate calculations of the consequences of mix are thus far unavailable. It is therefore planned to compensate for the effects of mix empirically—that is, it is planned to design and engineer for sufficient margin in ignition conditions and gain to compensate for whatever degradation the mix may cause.

The lack of a definitive, quantitative understanding of the origins and evolution of mixing has raised concerns that isolated bumps and defects in the capsule shell could give rise to spikes of wall material that would penetrate into the central fuel region. The potential for such an occurrence clearly is related to the precision of target fabrication; some target fabrication technology issues are discussed below.

Implosion Adiabat

Measurements indicate the existence of disparities between the calculated and actual adiabats on which NIF capsules implode. Some workers have postulated that the disparities are due to inaccuracies in tabulated plastic ablator (CH) release isentropes, but there appears to be no technical evidence to support this hypothesis.

LLNL briefings to the panel conveyed conviction that hot electron preheat from LPI in the NIF target has been adequately anticipated and that the implosion adiabat of the fuel can be managed by controlling shock heating. Nevertheless, the uncertainties concerning LPI processes within a target hohlraum (discussed below) and the strong sensitivity of a capsule's gain to preheat make the understanding and management of a capsule's implosion adiabat an area of concern to the panel.

Laser-Plasma Interactions

LPI diagnostics on an ID target assembly can only sample the small solid angle of light that is backscattered out of a hohlraum's laser entrance holes. The processes occurring inside the hohlraum, including those that can produce hot electrons, are difficult to observe. These circumstances significantly decrease the effectiveness of efforts to ascertain the adequacy of simulations of LPI.

Initial experiments on the OMEGA laser have shown disparities between modeling for both vacuum and gas-filled rugby hohlraums. Scattering of the inner beams entering a hohlraum is reported to be greater than predicted, providing specific evidence of simulation inadequacies.

Current simulations approximate LPI using inverse Bremsstrahlung energy deposition models in which the power balance of the beams is input by the user, although rad-hydro modeling has apparently been improved through the use of nonlocal electron transport models and detailed configuration analysis (DCA). Cross-beam transfer is estimated via analytic models. There is a fluid model for LPI, called PF3D, which includes approximate models of kinetic effects; the use of similar models might improve LPI simulations for laser fusion applications.

It appears to the panel that the current state of understanding and simulation capability of LPI presents a significant risk to both the NIC and the credibility of any indirect-drive IFE design concept, such as the Laser Inertial Fusion Energy (LIFE) initiative. The effects of LPI may be a central issue, contributing to observed disparities between measured and calculated implosion entropy, velocity, and shape in the NIC.

Capsule Fabrication

There is extensive experience in fabrication of NIC-style targets, and there is a high likelihood that the capsule and hohlraum system can be made to the desired specifications.

CONCLUSION 4-1: The national program to achieve ignition using indirect laser drive has several physics issues that must be resolved if it is to achieve ignition. At the time of this writing, the capsule/hohlraum performance in the experimental program, which is carried out at the NIF, has not achieved the compressions and neutron yields expected based on computer simulations. At present, these disparities are not well understood. While a number of hypotheses concerning the origins of the disparities have been put forth, it is apparent to the panel that the treatments of the detrimental effects of LPI in the target performance predictions are poorly validated and may be very inadequate. A much better understanding of LPI will be required of the ICF community.

CONCLUSION 4-2: Based on its analysis of the gaps in current understanding of target physics and the remaining disparities between simulations and experimental results, the panel assesses that ignition using laser indirect drive is not likely in the next several years.

The NIC plan—as the panel understands it—suggests that ignition is planned after the completion of a tuning program lasting 1-2 years that is presently under way and scheduled to conclude at the end of FY2012. While this success-oriented schedule remains possible, resolving the present issues and addressing any new challenges that might arise are likely to push the timetable for ignition to 2013-2014 or beyond.

CONCLUSION 4-3: Ignition of a laser-driven, indirect-drive capsule will provide opportunities for follow-up work to improve understanding of the potential for IFE.

- If ignition is achieved with indirect drive at the NIF, then an energy gain of 50-100 should be possible at a future facility. How high the gain at the NIF could be will be better understood by follow-on experiments once ignition is demonstrated. At this writing, there are too many unknowns to project a potential gain.
- Achieving ignition will validate the assumptions underlying theoretical predictions and simulations. This may allow a better appreciation of the sensitivities to parameters important to ignition.

USE OF LASER-DRIVEN, INDIRECT-DRIVE TARGETS IN A PROPOSED IFE SYSTEM

The proposed—and de facto—baseline model for a laser ID power plant is the LIFE initiative of LLNL. The discussions in this section are therefore based on that design as presented to the panel.

The current target design for LIFE was derived from the current baseline NIC design, with subtle but distinct differences. Modification was necessary to increase the calculated gain for IFE. Other modifications were to enable rapid, affordable fabrication in bulk, because the current plan for LIFE envisions firing approximately 1 million targets per day. The developers of LIFE plan to accommodate errors in the calculated target performance by adopting a design that is calculated to produce 125 percent of the gain for which LIFE was designed. The 25 percent surplus gain is viewed as a margin that would be eroded by the combined effects of inaccuracies in target design, fabrication, insertion, drive (shape, intensity, smoothing, and aiming), and LPI.

As discussed above, in evaluating the current NIC target, issues relating to the target implosion velocity, implosion symmetry, mix, the implosion adiabat, and LPI must be addressed. In spite of the modifications to the NIC target design that adapt it for use in LIFE, sufficient similarities persist that the preceding issues apply fully, unless and until optimization and other research conducted under the NIC program lead to a favorable resolution of the underlying uncertainties. The differences between the NIC and LIFE targets also raise additional issues, as discussed below.

Modifications to Increase Gain

The design approach to increasing the gain of the IFE capsule stems from an approximate analytical expression in which capsule yield is proportional to $E_{capsule}^{5/3}$, where $E_{capsule}$ is the energy absorbed by the capsule. The strategy is to increase the implosion energy primarily by increasing the drive temperature in the target hohlraum. The drive temperature is increased by increasing the laser driver energy and decreasing losses. The laser energy is to be increased from a maximum energy of 1.8 MJ at the NIF to 2.2 MJ for LIFE.

A hohlraum shaped like a rugby ball has been designed to more efficiently partition the drive energy; the redesign includes reducing the case-to-capsule diameter ratio to 2.0-2.4. The energy lost by reradiation from the hohlraum is to be reduced by the use of P2 LEH shields, and the conversion of absorbed energy to implosion energy is to be increased by using a high-density carbon (HDC) shell to increase the ablation efficiency. An illustration of the LIFE target design is shown in Figure 4-1.

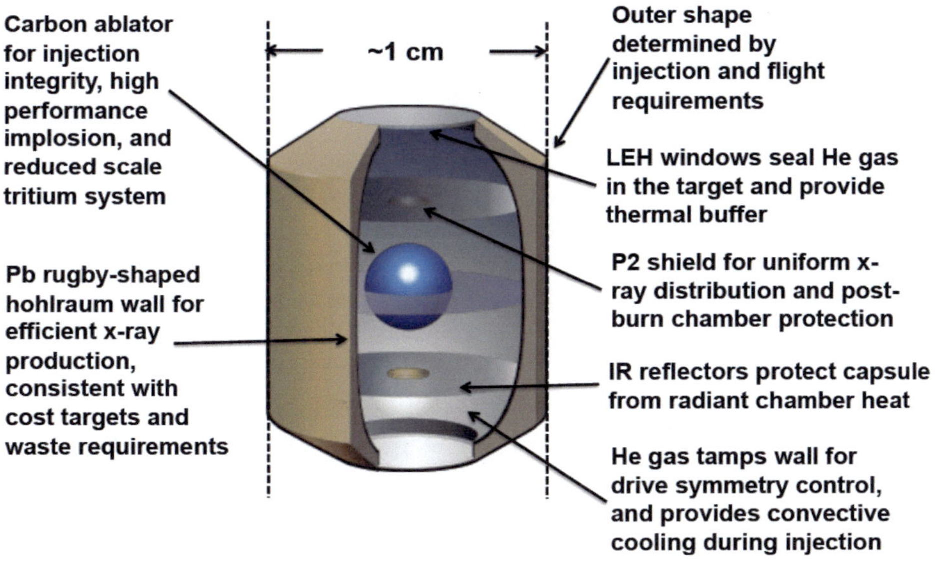

FIGURE 4-1 The LIFE target design. Modifications from the NIC target design include the curved ("rugby") inner wall of the hohlraum, the high-density carbon ablator, the LEH shields, and the P2 shine shields. SOURCE: M. Dunne, LLNL, presentation to the panel on July 7, 2011.

Modifications for Production Operation

The proposed manufacturing process of the LIFE target is a significant extension of the well-proven process for manufacturing targets for the NIC.

Capsule Fabrication

There is extensive experience in capsule fabrication, and it appears likely that the capsule can be made to the desired specifications. The technical challenges are (1) to demonstrate the formation of a uniformly thick, low-density (20 mg/cc) foam wall inside the diamond shell using a technique that is suitable for mass production and (2) to develop a cost-effective manufacturing process that can process more than 1 million targets per day through multiple steps where each target is individually handled. Proponents assert that automation can achieve the required throughput for an indeterminate capital and development cost; the bigger issue is whether the manufacturing can be done for the required per-item cost (estimated to be in the range of 20-40 cents).[4]

The method proposed for forming a uniformly thick fuel layer is a radical departure from the method used for making targets for the NIF. The reason for this new concept is to reduce the time required to form the fuel layer and thereby reduce the tritium inventory for the power plant. The design is for the fuel layer to be maintained as a supercooled liquid at a temperature sufficiently below the freezing point to achieve the required vapor pressure. The thickness uniformity of the fuel layer is expected to be provided by the 20 mg/cc CH foam wall, the interfacial liquid surface tension, and a controlled thermal profile along the surface of the hohlraum. This process has to be demonstrated. A critical technical milestone is to demonstrate that the DT liquid can be supercooled sufficiently to achieve the required vapor pressure, a property that has not been observed in cryogenic fluids.[5] A second technical challenge will be to preserve the uniformity of the liquid fuel when the capsule is accelerated to a velocity of 250 m/s into the target chamber. The low mechanical stiffness of the low-density foam and the low viscosity of the liquid will make the uniformity of the fuel layer thickness susceptible to the high acceleration loads.

Neither of the traditional methods of introducing fuel into the capsule—a capsule fill tube or diffusion filling—is feasible for power plant targets. A method would have to be developed to seal the capsules with a plug of some appropriate material after filling them with DT.

[4] D.T. Goodin, General Atomics, presentation to the main IFE committee on January 29, 2011.

[5] Different IFE target designs exist for different methods of achieving compression. Only one target design proposes supercooled DT liquid. If this step turns out to be physically impossible, then alternative designs will be explored.

Hohlraum

The rapid capsule insertion necessary for a power plant will require structurally rigid support for the capsule and the LEH shields. The hohlraum-capsule structure is a delicate and intricate design with tight assembly tolerances on how precisely the capsule needs to be positioned inside the hohlraum. In addition, there are two internal shine shields that need to be positioned precisely inside the hohlraum using a low-mass support structure so that neither the thermal profile nor the X-ray radiation flux within the hohlraum is excessively perturbed. Further work is required to define a construction that meets these requirements and will also survive the high acceleration loads experienced when the assembly is injected into the target chamber.

The hohlraum walls in the LIFE design are to be of a lead alloy that is optimized for high opacity at the capsule drive temperature. Current hohlraums are constructed either entirely of gold, or of gold-plated uranium. The latter are impractical for a high production rate. As an example, a firing rate of 10 Hz translates to 8.6×10^5 capsules fired per day. With a hohlraum mass of 3 g, 2.6 metric tons of lead must be collected and recycled per day. Using lead rather than solid gold will reduce both the start-up cost and the security requirements for the crucial processes of hohlraum material recycling and target fabrication.

Evaluation

In evaluating the current NIC target, issues relating to the target implosion velocity, implosion symmetry, mix, the implosion adiabat, and LPI were discussed above. The modifications to the NIC target design that adapt it for use in LIFE leave it fully vulnerable to the issues surrounding the performance of the NIC capsule, unless and until optimization and other research conducted under the NIC program lead to a favorable resolution of the underlying issues. The differences between the NIC and LIFE targets and drives also raise additional issues, which are discussed below. This section on the LIFE design concludes with an evaluation of the robustness of the LIFE target design.

Modifications to Increase Gain

The credibility of the effectiveness of the target design changes from NIC to LIFE is directly related to obtaining and understanding the desired performance of the NIC Rev 5.0 design and understanding its operation. The seriousness of the issues discussed in this section can be expected to become more apparent as the ignition campaign unfolds. Many of these changes are scheduled for study on OMEGA, the NIF, or both.

Capsule Implosion

The system modifications to increase the capsule drive are primarily intended to increase the energy of the imploding capsule; the implosion velocity is one indicator of this energy. The planned increase in the energy of the LIFE lasers should provide the most direct means of increasing the energy of an imploded capsule. The outlook for carrying out this plan is clearly independent of the target design, but any compromise in achieving this energy goal could severely reduce the likelihood of achieving sufficient gain for a power plant to be feasible.

Calculations indicate that a redesign of the target hohlraum from the cylinder shape used thus far at the NIF to a rugby shape can increase the drive temperatures for the enclosed capsule. However, initial experiments on the OMEGA laser using this hohlraum shape have shown disparities between the expected and measured temperatures. This trend was observed for both evacuated and gas-filled hohlraums. The disparities are not well understood and could be caused by increased importance of missing models of laser-plasma interactions or by something as simple as inadequate zone resolution. Although independent codes are used at the various laboratories, they tend to have similar models. Until a better understanding of the disparities between modeling and experiments on rugby hohlraums is achieved, there will be concerns that the needed drive temperatures might not be obtained.

Data appropriate for validating calculations of the temperature distribution and history in a rugby hohlraum are not yet in hand. Aspects of the calculations needing validation include the behavior of hohlraums with Pb walls, the radiation flow and hydrodynamic effects of P2 LEH shields, and the radiation hydrodynamics of a target utilizing a case diameter:capsule diameter ratio between 2.0 and 2.4. Such data must be acquired to attain confidence in predictions of target operation for LIFE.

Mix

The HDC to be used in the LIFE outer shell is a more complex material than the CH it is replacing; it exhibits a microcrystalline structure and is described by a complicated phase diagram. Because three-dimensional, directional irregularities are intrinsic to a microcrystalline structure, the potential for HDC to affect the hydrodynamic stability of the capsule requires further study.

LPI

The modifications of the LEH and the addition of the P2 shields to the NIC hohlraum create the potential for the LPI issues discussed above to be exacerbated by the use of a rugby hohlraum. Some increased effect could also be expected from the approximately 20 percent increase in laser power. The introduction of LEH shields with the rugby hohlraum may increase the mass of blown-off material in which LPI occur. Resulting changes in LPI phenomena may also change the implosion adiabat for the capsule.

Modifications for Production Operation

Target Fabrication

A target of this design has not yet been made, and new technologies will be required to make it. Only once the target is demonstrated to meet the specifications can the feasibility of mass-producing these targets for the desired cost be accurately assessed.

The plan to form the outer fuel layer of a LIFE target capsule by wicking liquid DT into a layer of nanoporous foam is a radical departure from the method used for making targets for the NIF. It will be necessary to demonstrate the formation of a uniformly thick, low-density (20 mg/cc) foam wall inside the HDC shell using a technique that is suitable for mass production. The efficacy of the planned smoothing mechanisms, as well as the ability to create and maintain the required thermal profile on the hohlraum through target insertion must also be demonstrated.

Other specific issues of concern include the need to eliminate the polishing step for the HDC shell and the significant length of time (approximately 2 days) involved for crucial manufacturing steps (chemical vapor deposition of the HDC and etching to remove the silicon mandrel) (Biener et al., 2009). The hohlraum-capsule structure is a delicate and intricate design with tight assembly tolerances on how precisely the capsule and two P2 LEH shields need to be positioned inside the hohlraum using low-mass support structures so that neither the thermal profile nor the X-ray radiation flux within the hohlraum is excessively perturbed. A construction method that meets these requirements is not yet available.

It would be important to the successful operation of the targets that the original specifications for the composition and uniformity of the lead mixture used to make the hohlraum walls be consistently maintained. The use of a "salted" Pb solution or alloy for the body of the target hohlraum would probably complicate the recycling process for that material. When it exits the reaction chamber, this material will have to be cycled through a full sequence of phases, proceeding rapidly from a solid to a plasma and then somewhat more slowly to a gas and a liquid. The composition of this liquid Pb mixture is unlikely to be uniform on the micron scale, and some portion of the other target components would also be present.

Whether fabrication to sufficiently tight specifications can be done for an acceptable per-item cost is an important question. It should be apparent from the discussion above that there are numerous technical challenges associated with developing an effective fabrication technology. However, the fuel costs for an inertial fusion power plant are much larger than is typical for the power industry,[6] so there is, financially, very little room for compromise. As currently envisioned, a viable technology must be capable of producing approximately 1 million targets a day through multiple steps in which each target is individually handled. Automation might achieve the required throughput by eliminating individual handling, but the associated capital and development costs are not known. The critical point from the standpoint of target design is that a compromise on any target specification or other aspect of fabrication quality would be likely to significantly reduce target gain.

Additional Considerations

The combination of extreme conditions that exist in a power plant reaction chamber and the very tight specifications that must be maintained for an IFE power plant to function result in an unusually tight coupling between the target design and some of what would typically be considered the separable engineering aspects of a power plant design. For the LIFE concept, the target insertion mechanism and the protection of the reaction chamber's laser windows fall into this category.

Target Insertion

The target must be positioned precisely at the desired location and in the desired alignment at the specified instant in time to uniformly drive the implosion. Positioning tolerance within approximately 1 cm of the optimum position was demonstrated as part of the High Average Power Laser (HAPL) program (see Box 4-2) using a smaller target than the proposed LIFE target. However, the conditions of the HAPL demonstration did not include transport through hot Xe gas, which will be present in the LIFE chamber to help protect the walls. Turbulence in this gas due to the ~10 Hz firing rate is inevitable, and its effect on target positioning is currently unknown. The LIFE targets are to be inserted into the reaction chamber in a manner that is most reminiscent of a bullet, requiring an acceleration of 400-500 *g* to reach the required 250 m/s velocity. This acceleration places very great demands on the technology for target fabrication.

The nominally low-mass supports for the P2 LEH shields and for the capsule itself must survive target acceleration with a sufficiently predictable geometry that their position satisfies tight specifications. It is even more important that the geometry of the capsule layers be as designed at shot time. The low mechanical stiffness

[6] The LIFE point design puts fuel costs at nearly 28 percent of the cost of electricity, about the same as the laser costs. From T. Anklam, LLNL, presention to the main IFE committee on January 31, 2011.

Box 4-2
Highlights of the High Average Power Laser Program

The goal of the HAPL program (FY1999-2009) was to pursue integrated development of science and technology for IFE that would be, to the extent possible, simple, durable, and affordable without sacrificing performance. The program featured parallel efforts on KrF and diode-pumped, solid-state lasers (DPSSLs). A high priority was placed on acquiring experimental data for both laser systems and technology concepts. The Sombrero Power Plant study (Sviatoslavsky et al., 1992) was used as a starting point.[1]

The HAPL program was based on laser-driven, direct-drive targets because of their potential for higher drive efficiency, simpler target fabrication, lower estimated cost, and smaller inventory for material recycling. Both conventional hot-spot ignition and shock ignition concepts were investigated. Predictions indicated that the drivers were equivalent for the conventional ignition and that the shorter-wavelength target produced higher gains for shock ignition. At the program goal of no more than 25 percent recirculating power, a combined driver target gain (*hG*) of 10 was needed, corresponding to a minimum target gain of 140 for a 7 percent efficient laser system (e.g., KrF). The HAPL program made significant progress in repetitive laser technologies for both diode-pumped Nd:glass and electron-beam-pumped KrF, demonstrating multihour runs at pulse rates from 5 to 10 Hz.

Research and development supported by the HAPL program included (1) calculations of neutron damage to optical ports and optics trains; (2) the development and successful testing of a new dielectric grazing incidence multilayer mirror for the first optical element of the laser system; (3) the development and demonstration of a method to mass-produce foam shells for target capsules; and (4) the development and demonstration of a cryogenic fluidized bed to make DT layers economically (the estimated cost of production was less than $0.17 each).

Target injection by both light-gas gun and magnetic slingshot was developed and tested. A method to improve capsule illumination accuracy detected the reflection ("glint"), from the moving capsule, of the light of a small laser to determine the target's trajectory. Real-time adjustment of the laser mirrors enabled illumination that was within 28 μ of the ideal to be demonstrated.

[1] An overview of the HAPL results is in Sethian et al., 2010.

of the low-density foam and the low viscosity of the DT liquid wicked into it may make it difficult to ensure a uniform thickness at shot time. These capabilities have not yet been demonstrated.

The HAPL program demonstrated active aiming of the drive laser that reduced its equivalent positioning error to 28 μ. The "glint" technique, in which the target capsule was illuminated during its trajectory through the essentially evacuated reaction chamber by a separate laser, utilized optical sensor location of the target by reflected laser light to determine the appropriate aim point. The firing rate in HAPL-sponsored tests was 5 Hz.

Successful translation of the glint technique to LIFE-style IFE would require that the target trajectory be sufficiently predictable to allow enough time to adjust the directions of the laser beam cones. Should perturbations of the target trajectory increase to problematic levels as it neared its aim point (the center of the turbulent region), very rapid detection and aiming adjustments would be needed to meet the 100 μ-equivalent error requirement for the LIFE design. Orientation of an ID target is also important, unlike the spherical HAPL target capsule. The target insertion technique includes inducing a spin along the LEH axis to stabilize its orientation. Successful irradiation would require that a target's angular momentum sufficiently overwhelm the effects of its hydrodynamic interaction with vorticity in the Xe fill of the reaction chamber that its orientation remains within acceptable bounds. Any second-order effects from also adjusting the aim of the laser beams are assumed here to be negligible. The difficulty of the other half of the glint technique—the illumination and detection of the target entering the reaction chamber—will be increased by the Xe fill. An assessment of this effect has not been presented to the panel.

Some unspecified portion of the gain margin calculated for the LIFE target has been allocated to compensating for nonoptimum insertion, but turbulence or other irregularities in the Xe gas through which the targets must pass could lead to sufficient inaccuracy not only to overwhelm that margin, but also to preclude capsule ignition. A key issue here is the repeatability of any phenomena that significantly perturb the target's trajectory.

The LEH shields are themselves inside LEH windows that are needed in the LIFE concept to separate the reaction chamber Xe from the He inside the hohlraum. The LEH windows also represent an interface between the cold interior of the target and the prevailing conditions of the reaction chamber. Some fraction of any Pb plasma or vapor from previous capsules through which a target travels might be expected to condense on the LEH windows during insertion and could affect the irradiation of the hohlraum interior.

Lastly, the accelerations must not cause any portion of the supercooled DT to change phase. Significant solidification would break the HDC ablator shell, and isolated solidification would create density nonuniformities that would spoil the implosion, either directly or by seeding hydrodynamic instabilities.

Target Robustness

The Merriam-Webster online dictionary[7] has several meanings for "robust," one of which is pertinent to the current discussion: "capable of performing without failure under a wide range of conditions." Robustness will be used in what follows to mean the quality of being robust according to this definition, with the regrettable caveat that the current state of the art limits an assessment's tie to reality to relatively indirect data. A result of this limitation is that degrees of robustness

[7] Available at www.merriam-webster.com.

actually indicate the assessed likelihood that a system can be made robust by actions and processes that are anticipated, proposed, or otherwise foreseeable, and, more fundamentally, the assessed likelihood that a system can be made to work at all.

Based on evaluations of the associated issues, the panel assesses the robustness of the physics design for the LIFE target concept to be low. The main factors leading to this assessment are the following:

- Ignition of a fusion target operating in the physics regime of laser-driven ICF has never been observed, but a robust design would have to reliably produce a large gain under much less controlled conditions than are normal in laboratory experiments. Moreover, the parameter space over which simulations predict adequate gain for the LIFE target capsule is relatively small, and the optimization of several parameters, an integral part of NIC, can be expected to further narrow the parameter space over which sufficient gain might be obtained;
- Significant departures from predicted operation have been observed on implosion experiments pertinent to the LIFE target design. These disparities, which were observed at both the NIF and the OMEGA lasers, relate directly to important aspects of target operation (e.g., implosion velocity), and the targets in which they were observed are the closest available analogues to the LIFE target. The discrepant data are important to the calibration or validation of the simulations on which predictions of the operation of the LIFE target are based, but tentative explanations of the disparities are at this time unsupported;
- To achieve the gain required for the LIFE plan to be viable, its target design incorporates modifications that are likely to further reduce the predictability of the target performance; and
- The outer, dense thermonuclear fuel region of the LIFE target is planned to be constructed of liquid DT wicked into low-density foam, but obtaining the gas pressure believed to be required for successful operation would require cooling the target capsule below the thermodynamic triple point for DT. The ability to create a LIFE target as currently designed therefore requires the existence of a physical phenomenon—the stabilization of a supercooled DT liquid in a low-density foam for an extended period of time—that has never been observed and for which there is no theoretical prediction.[8]

[8] There are studies that suggest it is possible to supercool hydrogen isotopes and other fluids (see, for example, Beaudoin et al., 1996). It remains unclear whether this effect can be achieved in the nanoporous hydrocarbon foam material, and if the corresponding vapor pressure is the desired value.

CONCLUSION 4-4: The target design for a proposed indirect-drive inertial fusion energy system (the Laser Inertial Fusion Energy, or LIFE, program developed by LLNL) incorporates plausible solutions to many technical problems, but the panel assesses that the robustness of the physics design for the LIFE target concept is low.

- The proposed LIFE target presented to the panel has several modifications relative to the target currently used in the NIC (e.g., rugby hohlraums, shine shields, and HDC ablators), and the effects of these modifications may not be trivial. For this reason, R&D and validation steps would still be needed.
- There is no evidence to indicate that the margin in the calculated target gain ensures either its ignition or sufficient gain for the LIFE target. If ignition is assumed, then the gain margin briefed to the panel, which ranged from 25 percent to almost 60 percent when based on a calculation that used hohlraum and fuel materials characteristic of the NIC rather than the LIFE target, is unlikely to compensate for the phenomena relegated to it—for example, the effects of mix—under any but the most extremely favorable eventuality. In addition, the tight coupling of LIFE to what can be tested on the NIF constrains the potential design space for laser-driven, indirect-drive IFE.

SOLID-STATE-LASER-DRIVEN, DIRECT-DRIVE FUSION

Current Status

The leader in direct-drive inertial confinement fusion with solid-state lasers is the Laboratory for Laser Energetics (LLE) at the University of Rochester, which operates the OMEGA Laser Facility (OMEGA and OMEGA EP) for the National Nuclear Security Administration (NNSA). LLE is conducting research into direct-drive ICF targets that utilize either the hot-spot ignition concept used by the NIC capsule or one of the more recent two-step ignition concepts (fast or shock ignition). The 60-beam OMEGA laser system, which delivers >30 kJ of 3ω light on target with 1-2 percent irradiation nonuniformity, has been operating since 1995, is fully instrumented, and is capable of up to 1,500 shots/year. The OMEGA EP laser system, which adds four NIF-like beamlines (6.5 kJ at 3ω), was completed in April 2008 and can propagate to either the OMEGA or OMEGA EP target chamber. Two EP beams can be operated as a high-energy petawatt (2.6 kJ in the infrared in 10 ps) system.

The current ICF program is aimed at exploring, understanding, and quantifying the physics issues of direct-drive laser targets at OMEGA drive energies and extrapolating the target performance to ignition and high-yield regimes. LLE has been routinely fielding cryogenic capsules since 2001 and has seen a steady improvement in implosion experiments as the quality of the ice layer and the centering of the target in the chamber have been improved. The flexible pulse-shaping capability of OMEGA enables the generation of multiple-picket pulse shapes that can drive ignition-scaled cryogenic DT implosions to ignition-relevant implosion velocities (3×10^7 cm/s) on a low adiabat ($\alpha \sim 2$-3[9]). The energies and relative timings of the three pickets and main pulse are adjusted to optimize the coalescence of four shocks to create a central hot spot, the same implosion strategy used at the NIF. Areal densities (ρr) up to 300 mg/cm^2 have been measured using a magnetic recoil spectrometer in cryogenic DT implosions on OMEGA drive at $\sim 8 \times 10^{14}$ W/cm^2 (Goncharov et al., 2010). The measured areal density in these experiments is larger than 88 percent of the predicted one-dimensional (1-D) value. The measured area mass density, ion temperature, and neutron yield can be combined with computed 1-D neutron yield to estimate the overall ignition parameter (χ)[10] for these experiments. These OMEGA cryogenic implosions have achieved an appreciable fraction (~3 percent) of the overall ignition parameter. The low inferred adiabats of these targets suggest that hot electron production from LPI and deposition into the fuel are within acceptable limits.

LLE has developed a 1 MJ symmetric, direct-drive NIF ignition design using a triple-picket pulse scaled to NIF laser parameters[11] that has a 1-D gain of ~50. Because direct drive has higher implosion efficiency than indirect drive, it is calculated to produce higher target gains, which should lead to lower laser cost.

No existing solid-state laser system in a direct-drive configuration presently has sufficient energy to demonstrate ignition. A multilaboratory workshop was held in 2001 whose purpose was not to preclude direct drive on the NIF (Meyerhofer, 2001). It was also agreed that the change board process would be used to ensure that future modifications did not preclude direct drive on the NIF. However, it is not clear that the final assembly procedure strictly adhered to this principle.

Reconfiguring the NIF to symmetric direct drive geometry represents the lowest target physics risk but the highest facility cost, and it would disrupt weapons physics experiments using hohlraums. As an alternative, LLE has identified a so-called "polar drive" (PD) geometry that allows direct-drive target performance to be studied at lower facility cost and minimal disruption of other experiments but

---

[9] α is a measure of the degree to which the actual adiabat of the implosion exceeds the ideal Fermi-degenerate adiabat (for which $\alpha = 1$).

[10] The ignition parameter is the energy that would have had to be absorbed by the target to produce ignition based on the other parameters achieved in the implosion—symmetry, density, and so on, as calculated in simulations.

[11] This involves targets whose dimensions are scaled down from the ignition design due to the reduced energy on OMEGA relative to the NIF.

at the price of higher target physics risk. Calculations predict that by repointing the beams from the existing laser ports, a uniform target drive can be achieved with PD irradiation, assuming that the irradiation at the equator is compensated by increased laser intensity. The risk is that the oblique irradiation at the equator occurs at lower densities, which reduces laser absorption and hydroefficiency and requires lateral heat flow to the equator from nonradial beams (Skupsky et al., 2004). The NIF triple-picket PD design with expected nonuniformities and multiple phase-modulation frequencies (multi-FM) beam smoothing achieves a calculated two-dimensional (2-D) gain of 32.

LLE has identified five changes on the NIF that would implement a PD capability for an ignition demonstration. OMEGA EP can be used to test many of the modifications, including multi-FM 1-D SSD beam smoothing,[12] and to validate laser performance. Advanced two-step ignition concepts such as shock ignition (SI) or fast ignition (FI) provide alternatives to conventional hot-spot ignition. If successful, these ignition options will open the path to high-gain ICF (G ~ 150) for ~1 MJ laser drivers (Perkins et al., 2009; Betti et al., 2006).

Fast ignition requires a combination of long-pulse (implosion) and short-pulse (FI) lasers. Aspects of FI both by electrons[13] and protons[14] were briefed to the panel. Integrated FI experiments have begun on OMEGA as part of the program of the Department of Energy (DOE) Office of Fusion Energy Sciences, which is studying the fast-electron coupling into a compressed core. The inferred laser-to-target heat coupling of ~3.5 percent needs to be increased significantly for FI to be a viable concept. Integrated simulations of electron-driven FI experiments are challenging and do not presently suggest ways of improving the target coupling. In principle, FI can also be achieved with protons accelerated by ultrashort-pulse lasers, which has the advantage of ballistic ion transport and sharper energy deposition. However, proton FI is hindered by lower laser conversion efficiency (~10 percent experimentally), a high intensity requirement (~10^{20} W/cm^2), and a high proton-dose requirement (~10^{16} protons) that complicates target fabrication. Further, a more complicated capsule design is required if a reentrant cone is used to protect the proton-generation foil. Although there is international interest in FI (e.g., the Fast Ignition Realization Experiment (FIRE) project at ILE/Osaka and HiPER in the United Kingdom), funding is presently insufficient for FI to challenge the mainline

[12] One-dimensional smoothing by spectral dispersion (SSD) with multiple phase-modulation frequencies (multi-FM) requires preconditioning the laser pulse with three high-frequency modulators to increase the bandwidth and is followed by a dispersion grating to increase the temporal skew. Multi-FM 1-D SSD has been optimized to provide the required beam smoothing to enable PD ignition. See Marozas et al., 2010.

[13] D. Meyerhofer, LLE, "Fast and Shock Ignition Research," presentation to the panel on July 6, 2011.

[14] J. Fernandez, LANL, "Inertial Confinement Fusion (ICF) Targets at Los Alamos National Laboratory," presentation to the panel on May 10, 2011.

programs on the NIF or the Laser Megajoule Facility (LMJ), which is under construction in France. Furthermore, the recently proposed concept of SI appears to be an easier and more attractive alternative to standard hot-spot ignition. SI utilizes a standard long-pulse laser beam with a pulse shape that provides a high-intensity spike at the end of the main drive pulse. The SI concept has been tested using CH shells on OMEGA. Higher areal densities (30 percent) and significantly higher neutron yields (~4×) were achieved with SI pulse shapes (Theobald et al., 2008).

Continued fundamental research into FI theory and experiments, the acceleration of electrons and ions by ultrashort-pulse lasers, and related high-intensity laser science is justified. However, issues related to low laser-target energy coupling, a complicated target design, and the existence of more promising concepts (such as SI), led the panel to the next conclusion on the relative priority of FI for fusion energy.

CONCLUSION 4-5: At this time, fast ignition appears to be a less promising approach for IFE than other ignition concepts.

Recent and Upcoming Work

The in-flight shell adiabat has been tuned by means of shock-velocity measurements using a variant of the NIF "key-hole target" (Boehly et al., 2011). Cross-beam energy transfer (CBET) has been identified as an issue that may be reducing laser energy absorption on OMEGA by 20 percent. Near-term experiments are planned to study mitigation strategies using modified phase-plate designs. Initial shock ignition designs for the NIF have 1-D gains of 70 at 680 kJ, with about half of that total energy in the shock generation pulse. PD diagnostic commissioning targets using existing ID phase plates are being imploded on the NIF (Cok et al., 2008).

LLE continues to demonstrate hydroequivalent scaling experiments on OMEGA to validate design codes that are then used for PD ignition calculations for the NIF. Upcoming experiments using targets with improved quality and reduced offset from the target chamber center are predicted to increase the χ from 3 percent of ignition to 5-6 percent, achieving the maximum credible performance for a 30-kJ driver.

LLE is developing a project execution plan (PEP) to demonstrate PD ignition on the NIF in 2017.

Evaluation and Discussion of Remaining R&D Challenges

Direct-drive, capsule-implosion data exist only at the 30 kJ level. The predicted hydroequivalent scaling requires validation at the MJ energy level, including issues of LPI, shock ignition at MJ energies, and symmetry. The modifications of the NIF for PD need to be developed and tested on OMEGA and deployed on the NIF. There are target physics risks for polar drive that need to be studied. Further, there are target fabrication, injection, and survival issues that are specific to the direct-drive approach. Specific issues are discussed individually below.

LPI

The larger energies for ignition targets are achieved through longer laser pulses, which result in long-scale-length plasmas that are more susceptible to LPI. There is a need to study and demonstrate acceptable laser energy deposition and hot electron production for ignition-scale plasmas. Relevant experiments can be done on OMEGA EP, which has NIF long-pulse beam lines. In particular, planar two-plasmon decay (TPD) experiments can quantify the hot electron production by collecting all electrons.

There are critical uncertainties in extrapolating TPD physics in planar geometry to the oblique irradiation geometry of the equatorial beams for NIF PD. Integrated TPD experiments on OMEGA will be very important in quantifying the production and deposition of hot-electron energy.

The plasma physics community requires a better understanding of CBET, including better theory and modeling, additional measurements, and tests of potential mitigation techniques.

The ability to model underdense plasma conditions is important for understanding LPI, since most LPI depend exponentially on electron density and temperature. Continued development of these models—including the effects of nonlocal transport—is important, especially for PD beam geometries.

Shock Ignition

Fully integrated 2-D point designs for the NIF PD shock ignition targets are required in order to plan for experimental campaigns on the NIF. Experiments need to continue on OMEGA to identify whether there are any LPI issues that are unique to the SI approach, especially in PD geometries. Experiments need to be done on OMEGA and later on the NIF to determine whether the hot-electron production by the high-intensity spike is acceptable for high-gain target performance. Calculations and experiments need to be performed to study the implementation of shock ignition pulses, including the trade-offs among laser beam parameters, illumination symmetry, and SI performance.

Symmetry

It remains to be seen whether sufficiently smooth laser beams can be created on the NIF to allow direct drive experiments, particularly in the PD geometry. Pointing errors and nonradial deposition geometries could lead to low-mode symmetry errors. Insufficient beam smoothing could lead to high-mode asymmetries. Symmetry issues related to providing both normal and high-intensity beams to illuminate SI targets need to be investigated, including calculations and experiments in PD geometry.

Reconfiguring the NIF for Polar Drive

These steps need to be taken to enable polar drive experiments on the NIF:

- Demonstrate new multi-FM 1-D SSD beam smoothing technique and validate on OMEGA EP.
- Design and demonstrate tailored phase plates to increase equatorial beam coupling.
- Design and demonstrate polarization smoothing for OMEGA EP to reduce focal-spot irradiance modulation. Design and demonstrate distributed polarization rotators (DPRs) that are sufficient to achieve polar-drive ignition on the NIF.
- Demonstrate integrated NIF PD beam smoothing on OMEGA EP.
- Complete development of a NIF fill-tube target that meets polar-drive ice layer specifications.
- Complete development of concepts for a PD ignition target insertion cryostat.

Polar-Drive Physics

Understanding of the following areas of polar-drive target physics need to be improved:

- Deposition in low-density plasma by oblique beams at equator, including three-dimensional (3-D) laser ray trace algorithms that are compatible with PD geometry.
- Ability of laser to deliver increased intensity to equatorial beams.
- Nonlocal transport and heat conduction for nonradial beams; this may require extensions to existing theory and algorithms.
- Possible LPI issues unique to PD illumination geometry; e.g., CBET between overlapping beams.

CONCLUSION 4-6: The prospects for ignition using laser direct drive have improved enough that it is now a plausible alternative to laser indirect drive for achieving ignition and for generating energy.

- The main concern with laser direct drive has been the difficulty of achieving the symmetry required to drive such targets. Advances in beam-smoothing and pulse-shaping appear to have lessened the risks of asymmetries. This assessment is supported by data from capsule implosions (performed at the University of Rochester's OMEGA laser), but it is limited by the relatively low drive energy of the implosion experiments that have thus far been possible. Because of this, the panel's assessment of laser-driven, direct drive targets is not qualitatively equivalent to that of laser-driven, indirect-drive targets.
- Further evaluation of the potential of laser direct-drive targets for IFE will require experiments at drive energies much closer to the ignition scale.
- Capsule implosions on OMEGA have established an initial scaling point that indicates the potential of direct-drive laser targets for ignition and high yield.
- Polar direct-drive targets will require testing on the NIF.
- Demonstration of polar-drive ignition on the NIF will be an important step toward an IFE program.
- If a program existed to reconfigure the NIF for polar drive, direct-drive experiments that address the ignition scale could be performed as early as 2017.

Potential for Use in an IFE System

If ignition and high yield can be demonstrated for DD targets, the higher target gain translates into greater system efficiency and lower laser energy (size). The even higher predicted gains of shock ignition targets make this DD concept very attractive. Shock ignition is not an option for ID targets because of the inherent integrating nature of the hohlraum, which limits the ability to spike the temperature drive.

Demonstrating PD ignition on the NIF is an important step toward an IFE program. This should include experiments to explore the performance of shock ignition targets on the NIF.

To date, the LLE ICF program has been focused on the development of laser-beam-smoothing technologies and single-shot ICF target physics experiments, which is the appropriate scope of the NNSA program. With the exception of some work in developing mass-production techniques for fabricating cryogenic DD targets and studying their survival in IFE-relevant thermal environments, LLE has not conducted research into either repetitive solid-state laser technologies or the host of issues associated with an IFE power plant. Through the HAPL program, LLNL has been the lead laboratory in developing repetitive solid-state lasers (DPSSL technology). Similarly, through the HAPL program, the Naval Research Laboratory (NRL) has supported the study of many of the technology and material issues related to the operation of a DD power plant. This suggests that there are opportunities for teaming among LLE, LLNL, and NRL if an IFE program is established to explore the potential of a DD fusion power plant with solid-state lasers. Further, LLE has much to contribute in target physics and target fabrication if KrF lasers prove more attractive as the laser driver in a DD power plant.

Additional Considerations

Target Injection

A key issue here is the repeatability of any phenomena that significantly perturb the target's trajectory.

Survival of Cryogenic Target

LLE has been studying the survival of cryogenic DD targets via complete Monte Carlo and computational fluid dynamics modeling of heat load to the target and its effect on the ice during injection into the chamber. These calculations will be supplemented by experiments in a surrogate IFE chamber. This issue was also addressed in the HAPL program, but more study is needed.

Reactor Chamber Issues

Most direct-drive IFE schemes are predicated on a dry-wall concept and an evacuated chamber. A host of structural and material issues need to be addressed. The HAPL program supported initial research in most of these areas, but much more work will be required before a power plant design can be completed. The HAPL final optic train was designed to meet the requirements for illumination uniformity, adequate tritium breeding, the threshold for damage to the grazing incidence metal mirror, and neutron damage to the conventional DD target. This design was applicable to both DPSSLs at 351 nm and KrF at 248 nm (Sethian et al., 2010).

CONCLUSION 4-7: In general, the science and engineering of manufacturing fusion targets for laser-based ICF are well advanced and meet the needs of those experiments, although additional technologies may be needed for IFE. Extrapolating this status to predict the success of manufacturing IFE targets is reasonable if the target is only slightly larger than the ICF target and the process is scalable.

However, subtle additions to the design of the ICF target to improve its performance (greater yield) and survivability in an IFE power plant may significantly affect the manufacturing paradigm.

CONCLUSION 4-8: There are important differences between the direct-drive and indirect-drive based targets. The direct-drive target is simpler to build than is the indirect-drive target, and it is more vulnerable to the environment when it is injected into the target chamber. Understanding these nuances and demonstrating a viable manufacturing process would likely be an important early priority for an IFE program because the quality and variability in the target's specifications can strongly affect the target's gain.

CONCLUSION 4-9: One major area where the IFE laser-driven target differs from the ICF target is the method of delivering the target to the target chamber at a high frequency. The high-velocity projectile techniques proposed for laser-based fusion show promise, but there has been little quantification of the degree to which the target will be compromised during the process and what effect any degradation may have on the target's gain. Also, changes that need to be made to the ICF target to improve its survivability in the IFE target chamber environment have been identified, but the consequence of these changes for the manufacturing process is not known. These are issues that need to be thoroughly addressed early in any future IFE program.

KRYPTON FLUORIDE LASER-DRIVEN, DIRECT-DRIVE FUSION

The leader in DD inertial confinement fusion with KrF lasers is NRL in Washington, D.C., which operates the Nike and Electra lasers. Nike is the world's largest KrF laser. Its amplifier with 60-cm aperture delivers a pulse of between 3 and 5 kJ at 248 nm to planar geometry targets using a smoothing technology called "induced spatial incoherence" (ISI). Nike has demonstrated "focal zooming," which allows the laser to more efficiently deliver late-time energy to the imploding spherical ICF pellet.

Electra is a repetitive KrF laser that was developed as part of the HAPL program to study the technology issues of repetition rate, durability, efficiency, and cost for inertial fusion energy. The HAPL program is discussed in Box 4-2. NRL has also developed the FAST (Gardner et al., 1998; Zalesak et al., 2005) radiation-hydrocode, which has several unique features that make it complementary to the ICF codes used at other laboratories.

The current ICF program on the Nike laser is focused on studying the hydrodynamic performance of planar targets accelerated by very smooth laser beams at 248 nm. LPI theories predict higher intensity thresholds for shorter wavelength lasers, proportional to the square of the wavelength. Further, shorter wavelengths enable higher absorption efficiency, larger drive pressure, and higher hydrodynamic efficiency. Experiments to quantify the growth of Richtmyer-Meshkov and Rayleigh-Taylor instabilities in planar cryogenic (deuterium wicked into foam) targets with thicknesses close to that of a high-gain target have been published (Pawley et al., 1999) and found to be in good agreement with theoretical predictions by the FAST3D code. The use of a thin high-Z layer to mitigate the imprinting of nonuniformities in the low-intensity laser foot was proposed and validated on Nike (Obenschain et al., 2002).

Further collaborative validation experiments on OMEGA demonstrated "significant and absolute (2X) improvements in neutron yield when the shells are coated with a very thin layer (~200–400 angstrom) of high-Z material such as palladium" (Mostovych et al., 2008). Thus, this imprint mitigation technique has been shown to work in both planar and spherical geometries at 248 and 351 nm. The utility of the high uniformity and higher ablation pressure generated by the Nike KrF laser was recently demonstrated in experiments on hypervelocity acceleration of planar targets in collaboration with researchers at the Institute of Laser Engineering at Osaka University in Japan. Whereas the Gekko XII/HiPER glass laser (351 nm) achieved a 700 km/s velocity, the KrF laser was able to achieve a 1,000 km/s foil velocity (Karasik et al., 2010). Extrapolating this performance to spherical DD implosions, ISI and zooming with a KrF laser offer the potential to use targets having lower aspect ratios and to reduce hydroinstability growth, thereby achieving higher target gain for less laser energy.

In 2008, Nike was upgraded to enable high-intensity LPI target experiments. The $2\omega_{pe}$ instability at quarter-critical density is of greatest concern in DD targets, where measurement of $\omega_o/2$, $3\omega_o/2$, and hard X-ray (>20 keV) emissions indicate the onset of the instability. The quarter-critical instability thresholds observed in Nike experiments with ISI-smoothed beams are in approximate agreement with planar beam $2\omega_{pe}$ theory, which does not account for the effects of beam smoothing, beam overlap, or saturated levels. This agreement includes an attempt to study the scaling with plasma scale length by varying the laser pulse length. OMEGA experiments with beams smoothed by SSD show similar agreement, and the predicted wavelength scaling appears to have been obtained. The OMEGA experiments have been modeled using the FAST and LILAC codes, both of which are in agreement with respect to the onset of LPI (Seka et al., 2009). However, DD ignition targets will likely need to operate above this theoretical threshold, and further research to understand, model, and measure LPI is required. This includes utilizing the NIF-equivalent OMEGA EP beam parameters to study LPI at plasma scale lengths that are relevant to ignition high-yield DD IFE targets.

A series of DD IFE target designs have been studied with the goal of maximizing target gain while minimizing laser energy. A conventional DD design provided

IFE-relevant 1-D gains (G ~ 100) at laser energies of ~1.3 MJ (Bodner et al., 2002). Later designs gave 1-D gains of order 50 with 500 kJ of KrF laser light by going to higher implosion velocities and using early-time spikes in the pulse shape to tailor the implosion adiabat and diminish Rayleigh-Taylor instability growth (Colombant et al., 2007).

The shock ignition concept proposed by Betti (Betti et al., 2007) and discussed in more detail in the preceding section, is now the baseline for KrF designs because of the higher predicted gains. An initial step in validating these designs was obtaining the agreement of FAST simulations of neutron yields with LLE simulations and experiments (Theobald et al., 2008). At IFE energies, FAST simulations of ISI-smoothed KrF beams using focal zooming give shock-ignition 1-D gains that are roughly twice as high as the best conventional designs (Schmitt et al., 2009). High-resolution, 2-D FAST simulations (for Legendre modes l = 1-256), which include the effects of inner and outer surface finishes and laser imprint, predict that these targets are robust to such perturbations.

The KrF research program would benefit from further 3-D implosion studies, improved LPI simulations, and experimental validation from LPI and implosion experiments on both OMEGA and the NIF in PD configuration. However, in PD geometry, the oblique irradiation near the equator occurs at lower densities, which reduces absorption and hydroefficiency and introduces nonradial beam illumination geometries and lateral heat flow. These are the remaining R&D challenges.

Recent and Upcoming Work

Having adequate numerical models for nonlocal thermal and hot-electron transport has been a challenge for several decades. Of special concern for DD, electron thermal transport in a laser-produced plasma cannot be described with a local approximation in many regions because the electron mean free path is longer than the temperature gradient scale length. NRL researchers have found that a Krook model provides reasonable descriptions of both preheat and flux limitation and have developed a computationally tractable algorithm; they are now verifying the accuracy of the model. This improved model will soon be available to apply to the analysis and design of ongoing experiments, as well as to the design of PD experiments on the NIF. These models are also relevant to the uncertainties in NIF hohlraum modeling.[15]

NRL has recently begun to simulate polar, DD implosions on the NIF using the FAST code. This will complement ongoing work by LLE in defining DD experiments for a polar-drive platform on the NIF. The growing collaboration will allow

[15] M. Rosen, LLNL, "Understanding of LPI and Its Impact on Indirect Drive," presentation to the panel on September 21, 2011.

development of conventional and shock ignition designs for the NIF and will enable use of the new Krook model to study the effect of nonlocal transport in the PD geometry.

Evaluation and Discussion of Remaining R&D Challenges

NRL presented a path forward to IFE DD target physics that included implosion experiments on OMEGA, LPI experiments on both Nike and OMEGA EP, and polar DD experiments on the NIF. The theory and simulation efforts included the development of better physics models for the FAST code, improved two- and three-dimensional hydroimplosion simulations, and improved ability to perform LPI simulations. NRL also proposed the development of one KrF IFE beam line that was capable of delivering ~20 kJ on target to study target interaction and LPI physics at IFE-relevant intensity and plasma scale lengths. The goal of this program, to be carried out in collaboration with LLE, would be to validate the fundamental physics of DD, to determine whether sufficient gains are feasible for IFE, and to validate the physics models for comparing DD target performance at 248 nm and at 351 nm.

The fundamental issues for DD capsules are the same at these two wavelengths, and the plans discussed in the solid-state laser DD section are all relevant and necessary. The importance of extending the OMEGA target performance database to NIF energies cannot be overemphasized. Specific issues relevant to the NRL program are discussed individually below.

Direct-Drive Theory and Physics Models

There is a continued need to develop improved physics models for DD in FAST, especially for potential megajoule-class experiments on the NIF, but in a nonradial, PD geometry. This includes continued development of nonlocal thermal and hot-electron transport models, improved nonlocal thermodynamic equilibrium (non-LTE) radiation modeling (particularly for thin, high-Z layers) and improved laser ray tracking for NIF PD geometries. There is also a need for improved LPI modeling, perhaps by teaming with other groups that have developed this capability and applying it at KrF wavelengths.

Laser-Plasma Interactions

As part of an increased effort toward understanding LPI, data on thresholds at KrF wavelengths will be useful. If a 20-kJ KrF laser was developed, it would provide the capability to study LPI at 248 nm in relevant scale-length plasmas and compare the results with OMEGA EP data. LLE is currently studying the role of CBET in DD experiments on OMEGA. KrF IFE designs may need to account for this physics, including the trade-off between CBET and illumination symmetry.

Polar-Drive Physics, Symmetry, and Shock Ignition

All of the issues listed under the solid-state DD section are relevant to the KrF DD program. Research into the physics issues of PD geometries, illumination symmetry in all DD geometries, and exploration of the potential of shock ignition as a high-gain target concept might best be pursued as a collaborative ICF/IFE program with both OMEGA and the NIF.

Capsule Fabrication, Injection, and Survival

These issues are similar to those already described for solid-state laser-driven targets.

Potential for Use in an IFE System

As noted in the preceding section, if ignition and high yield can be demonstrated for DD targets, the higher target gain translates into higher system efficiency and lower laser energy (size) at either 351 nm or 248 nm. If high-gain shock ignition proves feasible, the theoretical increase in gain for KrF with focal zooming as compared to frequency-tripled glass (with or without zooming) appears significant enough to merit serious consideration in IFE power plant economics. Further, from a driver perspective, the simplicity and effectiveness of ISI beam smoothing and focal zooming, the self-repairing nature of a gaseous gain medium, and the promising performance of the Electra laser system make KrF an IFE laser technology worth exploring. The final decision between 351 and 248 nm should be based on a total system performance analysis, including laser efficiency, durability, power plant integration issues, and overall target gain and performance. At this point, it would seem that an overall collaboration in direct-drive target physics and a competition between driver technologies at the beamline level would be a prudent technology maturation path.

CONCLUSION 4-10: Experiments on Nike in recent years give technical credence to using the deep-ultraviolet KrF wavelength to improve hydrodynamic coupling and increase LPI thresholds for direct-drive targets.

- Implosion experiments at 351 nm on OMEGA have made DD an attractive option for IFE. Planar experiments at 248 nm on Nike using ISI-smoothed beams have demonstrated the expected favorable scaling, with shorter wavelengths for laser absorption, increased drive pressure, and higher hydrodynamic efficiency, as well as higher LPI thresholds.

- The DD community would benefit from conventional and shock ignition experiments in PD geometry on OMEGA and the NIF, which might best be pursued as a national collaborative effort.
- Extending the Nike laser to 20 kJ would provide a valuable capability to study LPI and hydrodynamics at 248 nm in IFE-relevant scale-length plasmas and compare the results with OMEGA EP and NIF data.
- An overall collaboration in DD target physics and a competition between driver technologies at the beamline level would appear to be a prudent technology maturation path. The ultimate choice of laser wavelength and associated technology for DD IFE will be based on a total system analysis.

CONCLUSION 4-11: The lack of understanding surrounding LPI remains a substantial but as yet unquantified consideration in ICF and IFE target design.

RECOMMENDATION 4-1: DOE should foster collaboration among different research groups on the modeling and simulation of laser-plasma interactions.

HEAVY-ION-DRIVEN TARGETS

Current Status

The U.S. Heavy-Ion Fusion Science Virtual National Laboratory is a collaboration between LBNL, LLNL, and the Princeton Plasma Physics Laboratory (PPPL). The research is headquartered at LBNL. The Fusion Energy Sciences (FES) program within the DOE manages the heavy-ion fusion program. Historically, the mainline heavy-ion fusion (HIF) target design was developed to leverage the NIF experiments to demonstrate hot-spot ignition of an indirect drive target. Correspondingly, the most mature HIF target designs are for hohlraums with two-sided illumination (like the NIF) that indirectly drive a scale-up of the NIF capsule using repetitive accelerator technologies to provide the driver energy. ID hohlraums with NIF-like hot-spot ignition implosion physics are a well-documented approach (Callahan et al., 2002). For example, the 2002, two-dimensional Lasnex design (Zimmerman et al., 1978) called for a 7-MJ heavy-ion driver delivering 3- and 4-GeV Bi^{+1} ions to the hohlraum, giving a fusion gain of 68.

ID, and DD with hot-spot ignition or shock ignition using heavy-ion beams, are based on laser concepts but exploit the classical physics of ion-plasma energy deposition.[16] The briefing the panel received on heavy-ion target design at the July 2011 meeting[17] focused on the much newer X-target. The X-target is a HIF-motivated design that uses single-sided illumination by three sequential beam pulses and has features that offer new opportunities in accelerator driver technology, chamber technology, and driver-chamber interface.

Two preliminary target designs were presented to the panel at its Rochester meeting: (1) a 1-D Lasnex design of a DD target requiring 3 MJ of 3 GeV Hg^{+1} ions, giving a gain of ~150, and (2) a single-sided direct-drive X-target also utilizing 3 MJ of ions with a calculated 2-D gain of between 50 and 400 (see Figure 2-6). There are plans to extend the DD target design to 2-D design to incorporate a PD illumination geometry as well as a tamper and shock ignition assist.

Uranium beams of 80 GeV are already focused to <300 μm (full-width at half maximum) at GSI in Germany (transverse emittance sufficiently low), but beam current and space charge effects are small, and the bunch pulse durations are too long for fast ignition (>100 ns). Experiments at LBNL (NTX and NDCX-I) have shown that intense beam space charge can be neutralized with preformed target chamber plasma much greater than beam density. However, plasma neutralization cannot prevent the spread of the focal spot size due to chromatic aberrations (random momentum spread in the beam).

The sole LBNL target designer is continuing to evolve the X-target calculations in 2-D using the LLNL HYDRA code.

Evaluation and Discussion of Remaining R&D Challenges

The limitation of present accelerators in energy and focal intensity means that there are only a few data on ion-stopping powers in warm dense matter and no ICF target data. The PD and X-target performance estimates are purely based on rad-hydro code simulations that need to be greatly increased in sophistication and resolution to deal with all of the issues in a computational sense. The entry-level price of a heavy-ion target physics facility is sufficiently high that it is unlikely to be constructed by the DOE/NNSA program in the near or medium term.

Integrated 3-D Target Design

The 3-D nature of the HIF targets and highly sheared flows will require increasingly sophisticated simulations at very high resolutions (massively parallel).

---

[16] L.J. Perkins, LLNL, "Targets for Heavy Ion Fusion Energy," presentation to the panel on February 16, 2011.

[17] B.G. Logan, LBNL, "Heavy-Ion Target Design," presentation to the panel on July 7, 2011.

Mix

The sheared flows in the X-target with high-Z slide surfaces make mix with the DT fuel a serious concern.

Acceleration Compression Physics

It will be very challenging to reach the 200 ps/200 μm radius goals of the accelerator physics program. Ultimately, the limits of focusing and compression are determined by Liouville's theorem. The NDCX-II experiments will explore more intense beam compression and focusing physics related to subnanosecond heavy-ion shock ignition and fast ignition.

Neutralized Ballistic Focusing

The conceptual X-target designs assumed neutralized ballistic focusing of heavy ions through a background chamber plasma as simulated by the IBEAM systems code (Barton et al., 2005). Some panel members question the maturity of the models for dynamic charge state; the degree of neutralization in the reactor chamber environment; and the potential impact of beam space charge on the final focus. This is a transport issue that is unique to heavy-ion fusion and will require further research through detailed simulations and validation by experimental data (Sharp et al., 2004).

Potential for Use in an IFE System

All three heavy-ion target physics options are intended to use multiple-beam linac drivers with thick liquid-protected chambers to mitigate material neutron damage risks. The liquid-protected chamber technology is synergistic with some aspects of the pulsed-power approach to IFE.

In principle, the injection of targets into the reactor chamber for heavy ions has the same features as laser fusion. Light-gas-gun or magnetic-slingshot systems developed for laser fusion should be applicable. If the heavy-ion chamber uses a liquid lithium protection for the first wall, there may be some differences in injection system implementation and the specifics of cryogenic layer survivability in the reactor environment, which would be accounted for in a detailed system study.

All of the DD heavy-ion fusion target concepts are at a very early stage. Similarly, the proposed novel accelerator techniques for compressing heavy-ion beams to 200 ps with focusing to 200 μm radius are challenging and at an early stage of research. While heavy ions may represent a promising long-term option for efficient, reliable, repetitive fusion power plants, they probably represent a second- or third-generation capability.

CONCLUSION 4-12: The U.S. heavy-ion-driven fusion program is considering direct-drive and indirect-drive target concepts. There is also significant current work on advanced target designs.[18] This work is at a very early stage, but if successful may provide very high gain.

- The work in the HIF program involves solid and promising science.
- Work on heavy-ion drivers is complementary to the laser approaches to IFE and offers a long-term driver option for beam-driven targets.
- The HIF program relating to advanced target designs is in a very early stage and is unlikely to be ready for technical assessment in the near term.
- The development of driver technology will take several years, and the cost to build a significant accelerator driver facility for any target is likely to be very high.

Z-PINCH TARGETS

Description of Current U.S. Efforts

The main research in Z-pinch-driven ICF is performed at Sandia National Laboratories in Albuquerque, New Mexico. After the conversion of the PBFA-II accelerator to "Z" in 1997 to increase the radiated power from its wire-array Z-pinches, Sandia transitioned its ICF research from light-ion beam drivers to Z-pinches. The initial ICF concepts utilized thermal radiation from Z-pinches to indirectly drive ICF capsules. For example, the double-ended hohlraum concept drew heavily from ID ICF design experience at the NIF. Initial experiments on this concept demonstrated control of radiation symmetry via backlit capsule implosions; however, calculations showed that significant fusion experiments required much higher currents than achievable on Z (60 MA for high yield versus 20 MA Z capability). After completion of the Z Refurbishment Project in October 2007 (26 MA peak current), the NNSA issued guidance that the primary mission of Z should be to support the Science Campaigns within its Stockpile Stewardship program, especially in the areas of dynamic materials and nuclear weapons effects.

Presently, the limited portion of the Z experimental program that is devoted to ICF research is focused on concepts utilizing the DD of high magnetic field pressure to implode DT fuel to fusion conditions, citing an estimated 25-fold increase in theoretical efficiency for direct magnetic drive versus indirect X-ray drive. The Magnetized Liner Inertial Fusion (MagLIF) concept (see Figure 2-7) has been theoretically developed, and initial experiments to study the stability of the shell during magnetic implosion have been completed. Future experiments will add laser

[18] Advanced designs include DD, conical X-target configurations (see Chapter 2).

preheat to the magnetic implosions, with the eventual goal of $G = 1$ laboratory breakeven (DT fusion yield equals energy delivered to fuel). Quantitatively, this translates to ~100 kJ DT yields, although D_2 experiments will initially be performed for simplicity. High-yield (GJ-class), high-gain (>500) target designs are under development. Much of the relevant physics can be tested on Z.

R&D Challenges and Requirements

Some Z-pinch IFE system concepts were developed several years ago during a brief period when limited funding for IFE technology was provided within the NNSA ICF program. The concept of a recyclable transmission line (RTL) was explored as part of this technology project, although it was intended for use with the ID target designs that were being studied at that time. Extrapolated calculations of Z-pinch target designs typically require around 60 MA of current to be delivered from the pulsed-power driver to the implosion system to achieve high fusion yields. In contrast to laser and heavy-ion targets, which receive their energy from beams that are transported either in a vacuum or through small amounts of gas within the reactor chamber, the RTL directly connects the driver to the Z-pinch fusion target. This energy delivery strategy leads to a unique set of challenges and requirements for achieving the Z-pinch fusion system performance. The economics of this system design favor a low repetition rate and a high fusion target yield.

Technical and program managers at Sandia indicated to the panel that they perceive that ICF target research is not considered a high priority given the extensive funding necessary for the NIC and DOE's current prioritization of high-energy-density physics experiments on Z (e.g., the plutonium equation of state). Nevertheless, the existing program recently accommodated a modest amount of scientific work that shows significant promise for IFE. However, magnetically driven ICF ultimately needs to achieve robust fusion burn conditions, just as laser or heavy-ion ICF do. It has unique features that appear to the panel to provide an alternative risk-mitigating path to fusion energy. The Sandia Z100 program has been developed to address some of the key target physics issues in pulsed-power ICF. The pulsed-power technology program within the NNSA Science Campaigns is developing some of the next-generation technologies that would advance the pulsed-power driver issues of a fusion energy technology program. The following summarizes the overall program status:

- Single-shot, magnetically driven fusion target designs, funded by the NNSA, are being investigated on the Z accelerator.
- The MagLIF concept has been developed to exploit the favorable ignition requirements that, in theory, apply to target designs with magnetized and

preheated fuel. The MagLIF design is to be investigated in near-term validation experiments and simulations.

- Benchmark experiments on Z have shown excellent agreement between magneto-Rayleigh-Taylor simulations and observations.
- Development of an overall system for pulsed-power IFE was supported from 2004 to 2006 by modest (~$10 million) internal research funding. Sandia has indicated that internally funded research ($700,000) is now under way to continue the development of the RTLs.

Numerous issues surrounding target physics, driver technology, and fusion power system parameters stand between the current state of technology and magnetic IFE. These issues include the following:

- Liner dynamics
  —Obtain requisite velocities with suitable shell integrity.
  —Demonstrate sufficient control over the fuel adiabat during the implosion (e.g., pulse shaping).
  —Demonstrate tolerable levels of mixing at stagnation.
  —Demonstrate required level of axial asymmetry.
  —Demonstrate required level of azimuthal asymmetry.
- Fuel assembly
  —Demonstrate the required stagnation pressure.
  —Demonstrate required confinement time.
  —Compress sufficient current to a small radius to create extreme conditions.
  —Compress magnetic flux in the stagnating plasma.
- Driver scaling
  —Determine the driver parameters required for ignition and/or high yield.
  —Demonstrate scientific breakeven and support target approach with validated simulations.
  —Develop robust, high-yield targets designs in state-of-the-art 2-D and 3-D simulations.
  —Demonstrate a repetitive coupling with an RTL system.
  —Design a system for reliably creating, handling, and utilizing repetitive, high fusion yield with high availability.

Some additional specific technical issues still need to be explored:

- The MagLIF target design benefits from short implosion times; that is, the final density of the imploded fuel varies as (100 ns)/implosion time. However, the cost and the complexity of the pulsed-power driver have the opposite scaling. It was also stated that some target designs might be able to operate at longer implosion times. This would obviously be a huge lever arm on the total system that requires further investigation.

- The MagLIF performance scaling simulations have been primarily performed in 1-D, with limited exploration of 2-D Rayleigh-Taylor instability issues. However, the physics of thermal conduction and transport in magnetized plasmas is fully 3-D in nature and requires exploration in greater detail. 1-D simulations provide ideal energy scaling; 2-D begins to bring in Rayleigh-Taylor instabilities. Magnetized performance, however, will require 3-D studies.
- As stated by Sandia, "batch burn" (volume ignition) will result in a low yield, and a "levitated fuel" layer should give better performance. This will require additional calculations, target fabrication techniques, and experimental implementation. While providing improved performance, it also makes the fabrication and fielding logistics in a fusion power plant more complicated.
- Traditional magnetized target fusion concepts have not been shown to scale to high yield and gain. Sandia states that it has recently calculated high-yield performance with MagLIF targets. However, the additional cost of the magnets and optics that would be destroyed on each shot and the complexity of transporting the heater laser through the thick-liquid-wall chamber environment must both be accounted for in the system economics and design.
- References from the 2005 Sandia IFE program discuss potential issues of operating RTLs if the final radius and gap become too small. At that time the baseline power flow was relatively large wire-array Z-pinches. It will be important to study the compatibility of the RTL concept with the smaller diameter of direct magnetic-drive targets.

Potential for Use in an IFE System

Concepts for IFE systems using Z-pinch targets were presented to the panel,[19] but sufficient uncertainties remain that it would be premature to attempt an evaluation at this time. As presently envisioned, each 3-GJ fusion energy pulse would require the insertion, connection, and energizing of an RTL and fusion target assembly at a 0.1 Hz repetition rate. The assembly comprises an evacuated RTL system that contains the cryogenically cooled Z-pinch target at its center. The details of this concept are complex and will require extensive research and development if Z-pinches are pursued as an IFE technology. It is too early in both the target physics and fusion technology research programs to evaluate the target

[19] M. Cuneo et al., Sandia National Laboratories, "The Potential for a Z-pinch Fusion System for IFE," presentation to the panel on May 10, 2011.

fabrication and economic issues quantitatively, but the material and fabrication costs of the expended portions of the system will certainly be a factor in Z-pinch power plant economics. Because of the limited ICF target physics database, incomplete validation of the design tools and methodologies, and related lack of an integrated, high-yield target design, a consistent set of requirements and solutions for the pulsed-power driver, RTL, and ICF target cannot be articulated at this time. Therefore, the overall credibility of the energy delivery system and the ICF target performance cannot be quantitatively evaluated.

CONCLUSION 4-13: Sandia National Laboratories is leading a research effort on a Z-pinch scheme that has the potential to produce high gain with good energy efficiency, but concepts for an energy delivery system based on this driver are too immature to be evaluated at this time.

The Z-pinch scheme is completely different from the NIF and HIF approaches and therefore serves as risk mitigation for the ICF and IFE programs. It is not yet clear that the work at SNL will ultimately result in the high gain predicted by computer simulations, but initial results are promising and it is the panel's opinion that significant progress in the physics may be made in a year's time. The pulsed-power approach is unique in that its goal is to deliver a large amount of energy (~10 MJ) to targets with good efficiency (≥10 percent) and to generate large fusion yields at low repetition rates.

CONCLUSION 4-14: The target manufacturing and delivery processes that are proposed for direct-drive heavy-ion and pulsed-power fusion energy are less developed conceptually and technically than the targets for laser-based fusion energy. This is primarily because the priority has been to emphasize the implosion physics and driver issues (pulsed-power and linear accelerators). The pulsed-power target appears to be straightforward to manufacture, difficult to field, and challenging to reprocess after the thermonuclear event. In contrast, the heavy-ion targets possess many synergies with the laser-based target, but because a final target design is far from being defined, potential manufacturing complexities cannot be accurately assessed. The target delivery method for pulsed-power fusion is more conceptual than for laser- or heavy-ion-based fusion and presents very different problems—for example, a very much larger mass (~1,000 times larger), a slower replacement frequency (~100 times slower), and potentially a greater radioactive waste disposal problem.

OUTPUT SPECTRUM FROM VARIOUS IFE TARGETS

The fusion reaction of each type of IFE target produces a spectrum of threats (X-rays, ions, neutrons, and debris) to the first wall of the reaction chamber. The HAPL program studied the spectrum of threats to the first wall posed by direct-drive targets and developed candidate mitigation strategies and materials. It should be noted that while 14 MeV neutrons and 3.5 MeV α-particles are the universal products of the DT fusion reaction, the different target material and configurations for direct drive and indirect drive produce different threat spectra at the reactor chamber first wall. An IFE engineering test facility could be an intermediate step, before full-scale electrical power production, wherein fusion material issues could be studied.

Indirect Drive

The high-Z hohlraum materials used in ID absorb most of the α-particles and radiate more energy as X-rays. The actual threat spectrum is dependent on the details of the hohlraum design. For an ID, heavy-ion target, calculations show that 69 percent of the energy is in neutrons, 25 percent is in X-rays (500 eV peak), and 6 percent is in ions.[20] For the LIFE target, the X-ray fraction is about 12 percent, the ion fraction about 10 percent, and the remainder in neutrons.[21] X-rays are the dominant threat to the first wall for ID targets. The Osiris heavy-ion target chamber uses walls wetted by liquid lithium to mitigate the X-ray threat, while LIFE uses Xe gas to protect a dry solid wall.

Direct Drive

DD targets for both KrF and DPSSL systems produce the same threat spectrum, where approximately 1.3 percent of the energy is released in X-rays (4 keV peak) that produce surface deposition in less than the first 1 μm; 24 percent is in ions that have subsurface deposition in less than 5 μm, and the remainder is in neutrons that have volumetric deposition. Ions produce the greatest first wall heating for direct drive, and the implantation of α-particles presents a helium retention challenge. The HAPL program studied both of these challenges, combining modeling with experiments using lasers, ions, and plasma arc lamps to test thermomechanical cyclic stresses. The helium retention issue was similarly modeled, and experiments were performed on both the Van de Graff and the Inertial Electrostatic Confinement

[20] L.J. Perkins, LLNL, "Targets for Heavy Ion Fusion Energy," presentation to the panel on February 16, 2011.

[21] M. Dunne, LLNL, "LIFE Target System Performance," presentation to the panel on July 7, 2011.

fusion devices at the University of Wisconsin. A nanoengineered tungsten wall material showed an encouraging ability to mitigate helium retention. Experiments showed that cyclic heating in the IFE chamber mitigates helium retention.

Z-Pinch

The spectrum output issues associated with the RTL/Z-pinch system are unique to this approach. The mass of material in this assembly is much greater than in any other concept, leading to greater recycling requirements. Further, the interaction of the fusion output with the RTL structure could lead to unique problems with the formation of shrapnel and debris. These problems are not presently understood but appear to require a thick liquid-wall chamber.

TARGET FABRICATION

The primary concern of this panel with regard to ICF target fabrication relates to the technical feasibility of various proposed fabrication methods and the remaining technical risks and uncertainties associated with these methods. The question of whether the targets can be made cost-efficiently for a power plant is beyond the purview of this panel and is addressed by the National Research Council's IFE committee. Some promising approaches are discussed below.

Microfluidic Methodologies for Manufacturing Targets

The polymer shell that contains the DT fuel for DD laser and heavy-ion-beam fusion is proposed to be manufactured using a microfluidic droplet formation method.[22] This is an established technology that is used to make ICF capsules for current DD and ID experiments. The principle is to flow three immiscible fluids coaxially through two nozzles where the Rayleigh-Plateau instability that occurs in the region where they intersect produces individual droplets. Each droplet is an emulsion consisting of a thin shell of water surrounding a spherical oil droplet; these droplets are collectively immersed in oil. The thin shell of water contains the polymer precursors that form the plastic capsule. The final phase of the production process is to remove the fluids using supercritical drying.

This process has a very high production rate that is needed for a fusion energy program. However, the repeatability and precision of the process must be improved if the process is to be a viable option for an energy program. (The repeatability of the current process does not ensure that each capsule meets the required

[22] A. Nikroo, General Atomics, "Technical Feasibility of Target Manufacturing," presentation to the panel on July 8, 2011; see also Utada et al., 2007.

specifications, so each capsule is individually measured to determine its suitability; this raises the cost of the targets, which is acceptable for ICF experiments but not for an IFE program.) In all other aspects, this production process offers a potentially viable method for producing targets cost-effectively.

One modification to the current microfluidic method that may improve the reliability is to introduce electromechanical control into the process (Cho et al., 2003). This process, referred to as "lab-on-a-chip," has demonstrated the feasibility and benefits of using electric fields and electronics to control important steps in the target production process (Bei et al., 2010; Wang et al., 2011). This concept can potentially reduce the production time and physical size of a target production facility and address the precision and reliability concerns with the existing process. Further development of the process is needed.

The lab-on-a-chip concept is being evaluated as a method to accomplish the cryogenic operation of loading the DT fuel into the capsule.[23] Preliminary proof-of-concept experiments show that it is possible to form individual droplets of liquid deuterium of the correct size and wick them into a foam capsule in a short period of time. This would have the benefit of simplifying the target fueling process and shorten the process time, which would reduce the tritium inventory that is required by an IFE plant. Additional work is required to further develop this concept—specifically, to demonstrate that the process works with tritium and that it is practical to apply a condensed gas (argon, neon, or xenon) seal-coat onto the capsule once the fuel is loaded.

TWO OVERARCHING CONCLUSIONS AND A RECOMMENDATION

Based on the discussion in this chapter, the panel reached the following overarching conclusions and makes a recommendation:

OVERARCHING CONCLUSION 1: The NIF has the potential to support the development and further validation of physics and engineering models relevant to several IFE concepts, from indirect-drive hohlraum designs to polar direct-drive ICF and shock ignition.

- **In the near to intermediate term, the NIF is the only platform that can provide information relevant to a wide range of IFE concepts at ignition scale. Insofar as target physics is concerned, it is a modest step from NIF scale to IFE scale.**

[23] R. McCrory, LLE, "Target Fabrication for IFE Reactors: A Lab-on-a-Chip Methodology Suited for Mass-Production," submission to the panel on July 6, 2011.

- **Targets for all laser-driven IFE concepts (both direct-drive and indirect-drive) can be tested on the NIF. In particular, reliable target performance would need to be demonstrated before investments could confidently be made in the development of laser-driven IFE target designs.**

The NIF will also be helpful in evaluating indirectly driven, heavy-ion targets. It will be less helpful in gathering information relevant to current Z-pinch, heavy-ion direct-drive, and heavy-ion advanced target concepts.

OVERARCHING CONCLUSION 2: It would be advantageous to continue research on a range of IFE concepts, for two reasons:

- **The challenges involved in the current laser indirect-drive approach in the single-pulse NNSA program at the NIF have not yet been resolved, and**
- **The alternatives to laser indirect drive have technical promise to produce high gain.**

In particular, the panel concludes that laser direct drive is a viable concept to be pursued on the NIF. SNL's work on Z-pinch can serve to mitigate risk should the NIF not operate as expected. This work is at a very early stage but is highly complementary to the NIF approach, because none of the work being done at SNL relies on successful ignition at the NIF and key aspects of the target physics can be investigated on the existing Z-machine. Finally, emerging heavy-ion designs could be fruitful in the long term.

OVERARCHING RECOMMENDATION: The panel recommends against pursuing a down-select decision for IFE at this time, either for a specific concept such as LIFE or for a specific target type/driver combination.

Further research and development will be needed on indirect drive and other ICF concepts, even following successful ignition at the NIF, to determine the best path for IFE in the coming decades.

References

Barnard, J.J., R.O. Bangerter, E. Henestroza, I.D. Kaganovich, B.G. Logan, W.R. Meier, D.V. Rose, P. Santhanam, W.M. Sharp, D.R. Welch, and S.S. Yu. 2005. A final focus model for heavy-ion fusion driver codes. *Nuclear Instruments and Methods in Physics Research A* 544:243-254.

Beaudoin, G., P. Haljan, M. Paetkau, and J.R. Beamish. 1996. Freezing of molecular hydrogen and its isotopes in porous vycor glass. *Journal of Low-Temperature Physics* 105(1-2):113-131.

Bei, Z., T.B. Jones, and D.R. Harding. 2010. Electric field centering of double-emulsion droplets suspended in a density gradient. *Soft Matter* 6:2312-2320.

Belkov, S.A., S.G. Garanin, N.V. Jidkov, G.G. Kochemasov, and S.A. Suharev. 2010. The experimental studies conducted on the laser facilities of RFNC-VNIEFF: A review of the recent results. XXXVII Zvenigorod International Conference on Plasma Physics and Controlled Fusion.

Betti, R., A.A. Solodov, J.A. Delettrez, and C. Zhou. 2006. Gain curves for direct-drive fast-ignition at densities around 300 g/cc. *Physics of Plasmas* 13:100703.

Betti, R., C.D. Zhou, K.S. Anderson, L.J. Perkins, W. Theobald, and A.A. Solodov. 2007. Shock ignition of thermonuclear fuel with high areal density. *Physical Review Letters* 98:155001.

Biener, J., D.D. Ho, C. Wild, E. Woerner, M.M. Biener, B.S. El-dasher, D.G. Hicks, J.H. Eggert, P.M. Celliers, G.W. Collins, N.E. Teslich Jr,, B.J. Kozioziemski, S.W. Haan, and A.V. Hamza. 2009. Diamond spheres for inertial confinement fusion. *Nuclear Fusion* 49:112001.

Bobeica, M. 2009. Ph.D. Thesis. University of Rochester.

Bobeica, M., D.R Harding, and R.Q. Gram. 2005. An experimental method for measuring the response of a target to the thermal environment of the fusion reaction chamber. Twenty-first IEEE/NPS Symposium on Fusion Engineering.

Bodner, S.E., D.G. Colombant, A.J. Schmitt, J.H. Gardner, R.H. Lehmberg, and S.P. Obenschain. 2002. Overview of new high gain target design for a laser fusion power plant. *Fusion Engineering and Design* 60:93.

Boehly, T.R., V.N. Goncharov, W. Seka, M.A. Barrios, P.M. Celliers, D.G. Hicks, G.W. Collins, S.X. Hu, J.A. Marozas, and D.D. Meyerhofer. 2011. Velocity and timing of multiple spherically converging shock waves in liquid deuterium. *Physical Review Letters* 106(19):195005.

Callahan, D.A., M.C. Herrmann, and M. Tabak. 2002. Progress in heavy ion target capsule and hohlraum design. *Laser and Particle Beams* 20(3):405-410.

Carlson, L., M. Tillack, T. Lorentz, J. Spalding, N. Alexander, G. Flint, D Goodin, and R. Petzoldt. 2007. Target tracking and engagement for inertial fusion energy—A tabletop demonstration. *Fusion Science and Technology* 52(3):478.

Cho, S.K., H. Moon, and C.J. Kim. 2003. Creating, transporting, cutting, and merging liquid droplets by electrowetting based actuation for digital microfluidic circuits. *Journal of Microelectromechanical Systems* 12(1):70.

Cok, A.M., R.S. Craxton, and P.W. McKenty. 2008. Polar-drive designs for optimizing neutron yields on the National Ignition Facility. *Physics of Plasmas* 15:082705.

Colombant, D.G., A.J. Schmitt, S.P. Obenschain, S.T. Zalesak, A.L. Velikovich, J.W. Bates, D.E. Fyfe, J.H. Gardner, and W. Manheimer. 2007. Direct-drive laser target designs for sub-megajoule energies. *Physics of Plasmas* 14:056317.

Garanin, S.F., V.I. Mamyshev, and V.B. Yakubov. 2006. The MAGO system: Current status. *IEEE Transactions on Plasma Science* 34:2273-2278.

Gardner, J.H., A.J. Schmitt, J.P. Dahlburg, C.J. Pawley, S.E. Bodner, S.P. Obenschain, V. Serlin, and Y. Aglitskiy. 1998. Computational modeling of direct-drive fusion pellets and KrF-driven foil experiments. *Physics of Plasmas* 5:1935.

Goldston, R.J., and A. Glaser. 2011. Inertial confinement fusion energy R&D and nuclear proliferation: The need for direct and transparent review. *Bulletin of the Atomic Scientists* 67(3):59-66.

Goncharov, V.N., T.C. Sangster, T.R. Boehly, S.X. Hu, I.V. Igumenshchev, F.J. Marshall, R.L. McCrory, D.D. Meyerhofer, P.B. Radha, W. Seka, S. Skupsky, C. Stoeckl, D.T Casey, J.A. Frenje, and R.D. Petrasso. 2010. Demonstration of the highest deuterium-tritium areal density using multiple-picket cryogenic designs on OMEGA. *Physical Review Letters* 104:165001.

Jones, S.L., and F.N. von Hippel. 1998. The question of pure fusion explosions under the CTBT. *Science & Global Security* 7:129-150.

Karasik, M., J.L. Weaver, Y. Aglitskiy, T. Watari, Y. Arikawa, T. Sakaiya, J. Oh, A.L. Velikovich, S.T. Zalesak, J.W. Bates, S.P. Obenschain, A.J. Schmitt, M. Murakami, and H. Azechiet. 2010. Acceleration to high velocities and heating by impact using Nike KrF laser. *Physics of Plasmas* 17:056317.

Kirillov, G.A., G.G. Kochemasov, A.V. Bessarab, S.G. Garanin, L.S. Mkhitarian, V.M. Murugov, S.A. Sukharev and N.V. Zhidkov. 2000. Status of laser fusion research at VNIIEF. *Laser and Particle Beams* 18(2):219-228.

Kodama, R., H. Shiraga, K. Shigemori, Y. Toyama, S. Fujioka, H. Azechi, H. Fujita, H. Habara, T. Hall, Y. Izawa, T. Jitsuno, Y. Kitagawa, K.M. Krushelnick, K.L. Lancaster, K. Mima, K. Nagai, M. Nakai, H. Nishimura, T. Norimatsu, P.A. Norreys, S. Sakabe, K.A. Tanaka, A. Youssef, M. Zepf, and T. Yamanaka. 2002. Nuclear fusion: Fast heating scalable to laser fusion ignition. *Nature* 418:933.

Lindemuth, I.R., R.E. Reinovsky, R.E. Chrien, J.M. Christian, C.A. Ekdahl, J.H. Goforth, R.C. Haight, G. Idzorek, N.S. King, R.C. Kirkpatrick, R.E. Larson, G.L. Morgan, B.W. Olinger, H. Oona, P.T. Sheehey, J.S. Shlachter, R.C. Smith, L.R. Veeser, B.J. Warthen, S.M. Younger, V.K. Chernyshev, V.N. Mokhov, A.N. Demin, Y.N. Dolin, S.F. Garanin, V.A. Ivanov, V.P. Korchagin, O.D. Mikhailov, I.V. Morozov, S.V. Pak, E.S. Pavlovskii, N.Y. Seleznev, A.N. Skobelev, G.I. Volkov, and V.A. Yakubov. 1995. Target plasma formation for magnetic compression/magnetized target fusion (MAGO/MTF). *Physical Review Letters* 75(10):1953-1956.

Liu, C.S., and M.N. Rosenbluth. 1976. Parametric decay of electromagnetic waves into two plasmons and its consequences. *Physics of Fluids* 19(7):967-971.

Marozas, J.A., J.D. Zuegel, and T.J.B. Collins. 2010. Smoothing by spectral dispersion (SSD) for multiple-picket pulses on OMEGA and the NIF. *Bulletin of the American Physical Society* 55:294.

Massard, T. 2010. ICF in France, status and perspective. Fusion Power Associates 31st Annual Meeting and Symposium, Washington, D.C., December 1-2.

Meserve, R.A. 2011. Nuclear energy and climate change. *Nuclear Plant Journal* 29(3):24.

Meyerhofer, D.D. 2001. Direct drive ignition research. Presentation to the Campaign Review Meeting, Los Alamos, N.M., December 3-6.

Mostovych, A.N., D.G. Colombant, M. Karasik, J.P. Knauer, A.J. Schmitt, and J.L. Weaver. 2008. Enhanced direct-drive implosions with thin high-*Z* ablation layers. *Physical Review Letters* 100:075002.

Norimatsu, T., K. Nagai, T. Takeda, K. Mima, and T. Yamanaka. 2003. Update for the drag force on an injected pellet and target fabrication for inertial fusion. *Fusion Science and Technology* 43:339.

NRC (National Research Council). 1986. Review of the Department of Energy's Inertial Confinement Fusion Program. Washington, D.C.: National Academy Press.

NRC. 1990. Review of the Department of Energy's Inertial Confinement Fusion Program. Washington, D.C.: National Academy Press.

NRC. 1997. Review of the Department of Energy's Inertial Confinement Fusion Program: The National Ignition Facility. Washington, D.C.: National Academy Press.

Obenschain, S.P., D.G. Colombant, M. Karasik, C.J. Pawley, V. Serlin, A.J. Schmitt, J.L. Weaver, J.H. Gardner, L. Phillips, Y. Aglitskiy, Y. Chan, J.P. Dahlburg, and M. Klapisch. 2002. Effects of thin high-Z layers on the hydrodynamics of laser-accelerated plastic targets. *Physics of Plasmas* 9:2234.

Paine, C., and M.G. McKinzie. 1998. Does the U.S. science-based Stockpile Stewardship Program pose a proliferation threat? *Science and Global Security* 7:151-193.

Pawley, C.J., S.E. Bodner, J.P. Dahlburg, S.P. Obenschain, A.J. Schmitt, J.D. Sethian, C.A. Sullivan, J.H. Gardner, Y. Aglitskiy, Y. Chan, and T. Lehecka. 1999. Observation of Rayleigh–Taylor growth to short wavelengths on Nike. *Physics of Plasmas* 6:565.

Perkins, L.J., R. Betti, K.N. LaFortune, and W.H. Williams. 2009. Shock ignition: A new approach to high gain inertial confinement fusion on the National Ignition Facility. *Physcal Review Letters* 103:045004.

Petzoldt, R.W., D.T. Goodin, A. Nikroo, E. Stephens, N. Siegel, N.B. Alexander, A.R. Raffray, T.K. Mau, M. Tillack, F. Najmabadi, S.I. Krasheninnikov, and R. Gallix. 2002. Direct drive target survival during injection in an inertial fusion energy power plant. *Nuclear Fusion* 42:1351.

Schmitt, A.J., J.W. Bates, S.P. Obenschain, S.T. Zalesak, D.E. Fyfe, and R. Betti. 2009. Direct drive fusion energy shock ignition designs for sub-MJ lasers. *Fusion Science and Technology* 56:377.

Seka, W., D.H. Edgell, J.F. Myatt, A.V. Maximov, and H.A. Baldis. 2009. Two-plasmon-decay instability in direct-drive inertial confinement fusion experiments. *Physics of Plasmas* 16:52701.

Sethian, J.D., D.G. Colombant, J.L. Giuliani, R.H. Lehmberg, M.C. Myers, et al. 2010. The science and technologies for fusion energy with lasers and direct-drive targets. *IEEE Transactions on Plasma Science* 38(4):690-703.

Sharp, W.M., D.A. Callahan, M. Tabak, S.S. Yu, P.F. Peterson, D.V. Rose, and D.R. Welch. 2004. Chamber-transport simulation results for heavy-ion fusion drivers. *Nuclear Fusion* 44:S221.

Skupsky, S., J.A. Marozas, R.S. Craxton, R. Betti, T.J.B. Collins, J.A. Delettrez, V.N. Goncharov, P.W. McKenty, P.B. Radha, T.R. Boehly, J.P. Knauer, F.J. Marshall, D.R. Harding, J.D. Kilkenny, D.D. Meyerhofer, T.C. Sangster, and R.L. McCrory. 2004. Polar direct drive on the National Ignition Facility. *Physics of Plasmas* 11:2763.

Slutz, S.A., and R.A. Vesey. 2012. High-gain magnetized inertial fusion. *Physical Review Letters* 108(2):025003.

Sviatoslavsky, I.N., M.E. Sawan, R.R. Peterson, G.L. Kulcinski, J.J. MacFarlane, L.J. Wittenberg, H.Y. Khater, E.A. Mogahed, and S.C. Rutledge. 1992. A KrF laser driven inertial fusion reactor "SOMBRERO." *Fusion Technology* 21:1470.

Theobald, W., R. Betti, C. Stoeckl, K.S. Anderson, J.A. Delettrez, V. Yu. Glebov, V.N. Goncharov, F.J. Marshall, D.N. Maywar, R.L. McCrory, D.D. Meyerhofer, P.B. Radha, T.C. Sangster, W. Seka, D. Shvarts, V.A. Smalyuk, A.A. Solodov, B. Yaakobi, C.D. Zhou, J.A. Frenje, C.K. Li, F.H. Séguin, R.D. Petrasso, and L.J. Perkins. 2008. Initial experiments on the shock-ignition inertial confinement fusion concept. *Physics of Plasmas* 15:056306.

Tillack, M., A.R. Raffray, F. Najmabadi, and L.C. Carlson. 2010. Target and chamber technologies for direct-drive laser-IFE. Final Report for the IAEA CRP: Pathways to Energy from Inertial Fusion (IFE)—An Integrated Approach. UCSD-CER-10-01. University of California, San Diego.

U.S. DOE (U.S. Department of Energy), Office of Declassification. 2001. Restricted Data Declassification Decisions, 1946 to the Present (RDD-7). Item V.C.1.g (declassification action 98-15). January 1.

U.S. DOE, Office of Arms Control and Nonproliferation. 1995. The National Ignition Facility (NIF) and the Issue of Nonproliferation: Final Study. December 19.

Utada, A.S., L.-Y. Chu, A. Fernandez-Nieves, D.R. Link, C. Holtze, and D.A. Weitz. 2007. Dripping, jetting, drops, and wetting: The magic of microfluidics. *MRS Bulletin* 32:702-708.

Velikhov, E. 2008. 80 Years of Fusion. AAAS 2008 Annual Meeting. Boston, Mass. February 14-18.

Wang, W., T.B. Jones, and D.R. Harding. 2011. On-chip double emulsion droplet assembly using electrowetting-on-dielectric and dielectrophoresis. *Fusion Science and Technology* 59:240-249.

Zalesak, S.T., A.J. Schmitt, A.L. Velikovich, and J.H. Gardner. 2005. Modeling fluid instabilities in inertial confinement fusion hydrodynamics codes. *Physics of Plasmas* 12:056311.

Zimmerman, G.G., D. Kershaw, D. Bailey, and J. Harte. 1978. LASNEX code for inertial confinement fusion (A). *Journal of the Optical Society of America* 68:549.

Appendixes

A

Biographical Sketches of Panel Members

John F. Ahearne (NAE), *Chair,* is the executive director emeritus of Sigma Xi, The Scientific Research Society, an adjunct professor of engineering at Duke University, and an adjunct scholar at Resources for the Future. He has extensive expertise in nuclear and radiation engineering and risk assessment. His professional interests are in reactor safety, energy issues, resource allocation, and public policy management. Dr. Ahearne served in the U.S. Air Force from 1959 to 1970, resigning as a major. He has also served as deputy and principal deputy assistant secretary of defense (1972-1977), in the White House Energy Office (1977), as deputy assistant secretary of energy (1977-1978), and as commissioner and chairman of the U.S. Nuclear Regulatory Commission (chairman, 1979-1981). He is a fellow of the American Physical Society, the Society for Risk Analysis, the American Association for the Advancement of Science, the American Academy of Arts and Sciences, and a member of the National Academy of Engineering, Sigma Xi, and the American Nuclear Society. He has previously chaired or served as a member on committees for more than 30 other NRC studies. Dr. Ahearne received a Ph.D. in physics from Princeton University.

Douglas Eardley, *Vice Chair,* is professor of physics at the Kavli Institute for Theoretical Physics at the University of California, Santa Barbara. Dr. Eardley's research interests include general relativity: black holes, gravity waves, and quantum gravity; theoretical astrophysics: X-ray sources, quasars, active galactic nuclei, and cosmology; mathematical physics: nonlinear partial differential equations and geometry; and physics and society: national security, nuclear weapons, and arms control. Dr. Eardley has been a member of several National Research Council study committees, including the Working Group on Related Areas of Science of the Astronomy Survey Committee ("Field Committee") in 1979-1980; the Committee on the Atmospheric Effects of Nuclear Explosions in 1983-1984; and the Science Panel of the Astronomy Survey Committee in 1989-1990. He was chair of the External Advisory Board of the Institute for Fundamental Theory of the University of Florida at Gainesville from 1990 to 1994; a member of the Physics Advisory Committee of Lawrence Livermore National Laboratory from 1991 to 1996; the plenary speaker at the Texas Symposium on Relativistic Astrophysics in 1992; a member of the Openness Advisory Panel of the Secretary of Energy Advisory Board for DOE from 1996 to 2002; and co-coordinator of the Institute for Theoretical Physics' Program in Black Hole Astrophysics from 1999 to 2002. Professor Eardley has been a member of the JASON Study Group since 1981; a member of the National Security Panel of the University of California's President's Council on the National Laboratories from 2000 to 2007; chair of the External Review Panel for the Radiation Effects Sciences Program for Sandia National Laboratories since 2000; and a member of the Joint Mission Committee for Los Alamos National Laboratory and Lawrence Livermore National Laboratory since 2007. He received a B.S. in physics from the California Institute of Technology and M.S. and Ph.D. degrees in physics from the University of California, Berkeley.

Robert C. Dynes (NAS) is professor emeritus of physics at the University of California, San Diego. He served as the 18th president of the University of California (UC) from 2003 to 2007 and as chancellor of UC San Diego from 1996 to 2003. His position as chancellor followed 6 years in the physics department, where he founded an interdisciplinary laboratory in which chemists, electrical engineers, and private industry researchers investigated the properties of metals, semiconductors, and superconductors. Prior to joining the UC faculty, he had a 22-year career at AT&T Bell Laboratories, where he served as department head of semiconductor and material physics research and director of chemical physics research. Dr. Dynes received the 1990 Fritz London Award in Low Temperature Physics, was elected to the National Academy of Sciences in 1989, and is a fellow of the American Physical Society, the Canadian Institute for Advanced Research, and the American Academy of Arts and Sciences. He serves on the executive committee of the U.S. Council on Competitiveness. A native of London, Ontario, Canada, and a naturalized U.S. citizen, Dr. Dynes holds a bachelor's degree in mathematics and physics and an honorary doctor of law degree from the University of Western Ontario, and master's and doctoral degrees in physics and an honorary doctor of science degree from McMaster University. He also holds an honorary doctorate from Université de Montréal.

David Harding is a senior scientist at the University of Rochester's Laboratory for Laser Energetics and a professor in the Department of Chemical Engineering. His research interests include the science and engineering associated with the making of fuel capsules for fusion experiments performed at the University of Rochester's Laboratory for Laser Energetics. He has worked at the University of Rochester for 15 years; prior to that he was a senior research engineer in the Materials and Structures Division at the NASA Lewis Research Center. He has participated as a panel member on two review committees: the National Ignition Facility Target Fabrication Review (2008) at Lawrence Livermore National Laboratory and a DOE review of its Solar Thermal Program (1992). Dr. Harding received a Ph.D. from Cambridge University.

Thomas Mehlhorn is superintendent of the Naval Research Laboratory (NRL) Plasma Physics Division, and a member of the Department of the Navy Senior Executive Service with responsibility for a broad spectrum of research programs in plasma physics, laboratory discharge and space plasmas, intense electron and ion beams and photon sources, atomic physics, pulsed-power sources, radiation hydrodynamics, high-power microwaves, laser physics, advanced spectral diagnostics, and nonlinear systems. He began his career at Sandia National Laboratories in 1978 and worked on a variety of projects related to the generation, focusing, and interaction of intense beams of electrons and ions with plasmas. From 1989 to 1998 he was a manager in the Sandia Light Ion ICF Program, and from 1998 to 2006 he managed Sandia's High Energy Density Physics and ICF Target Design Department in the Pulsed Power Fusion Program. From 2006 to 2009 he was a senior manager with accountability for dynamic materials and shock physics, high energy density physics theory and modeling, and advanced radiographic source development and applications. Dr. Mehlhorn joined NRL in 2009. He is a recipient of two NNSA Defense Programs Award of Excellence (2007 and 2008), a Lockheed Martin NOVA award (2004), and an Alan Berman Research Publication Award from NRL (1983). Dr. Mehlhorn is a fellow of the American Association for the Advancement of Science (AAAS) in physics (2006). He serves on the Advisory Board for Plasma and Atomic Physics at GSI, Darmstadt, Germany (2004-present, chair in 2006). He is a member of the Nuclear Engineering and Radiological Sciences Department Advisory Board at the University of Michigan (1996-1999 and 2004-present), as well as of the University of Michigan College of Engineering Alumni Society board of governors (2009-present). In 2010 Dr. Mehlhorn served on the Department of the Navy Space Experiments Review Board as well as the University of Missouri's Research and Development Advisory Board. Dr. Mehlhorn received B.S., M.S., and Ph.D. degrees in nuclear engineering from the University of Michigan.

Merri Wood-Schultz is a part-time consultant for SAIC and serves as a laboratory associate at LANL for improvised and foreign devices. Dr. Wood-Schultz's early career focused on the physics design of secondaries of thermonuclear weapons. She was responsible for the conceptual and physics design of numerous nuclear tests and add-on experiments; the areas of focus of these tests included stockpile systems, weapons physics, and advanced development. Dr. Wood-Schultz played an active role in the development of nuclear weapons-related laboratory experiments (AGEX), serving as the lead designer for a series of experiments on the Sandia National Laboratories' SATURN pulsed-power machine and as a member of the inaugural LANCE (neutron scattering facility) Users Group. Later phases of Dr. Wood-Schultz's career included involvement in developing concepts and methods for certification without nuclear testing, notably the quantification of margins and uncertainty (QMU), and an increase in her work in nuclear intelligence. The latter led to a 6-month, change-of-station assignment to a DOE intelligence organization. Dr. Wood-Schultz is currently a member of the Nuclear Forensics Science Panel for the Department of Homeland Security and engages in continuing technical collaborations on nuclear weapons design, yield certification using QMU, and nuclear intelligence. Dr. Wood-Schultz became a fellow of Los Alamos National Laboratory in 2001, received the Department of Energy Award of Excellence in 1988, 1999, and 2004, the STRATCOM Medal of Excellence in 1997, and the Los Alamos National Laboratory Distinguished Performance Award in 1996. Dr. Schultz received B.S., M.S., and Ph.D. degrees in physics from the Georgia Institute of Technology.

George Zimmerman is a part-time consultant on computations and modeling for LLNL and on nuclear reactor modeling for TerraPower, LLC. He joined LLNL in 1970 as a staff member in the A Division, where he developed the LASNEX computer program to design laser fusion targets and analyze experiments. In 1980 he was appointed associate division leader in the X Division, where he led a group of physicists responsible for developing numerical methods to accurately perform integrated simulations involving laser absorption, magnetohydrodynamics, atomic physics, and the transport of photons, neutrons, and charged particles. From 1984 to 1987 he was leader of the Computational Physics Division. He then led the inertial confinement fusion code development project in the AX Division until his retirement. Mr. Zimmerman received the Department of Energy's 1983 E.O. Lawrence Award for contributions to national security and the 1997 Edward Teller Award for developing the LASNEX inertial confinement fusion code. He also received the Defense Programs Award of Excellence for significant contributions to the Stockpile Stewardship Program in 2002 and 2005. He retired from LLNL in 2007 and is currently a fellow of the American Physical Society. Mr. Zimmerman received a B.S. in physics from Harvey Mudd College and an M.A. in astronomy from the University of California, Berkeley.

B

Panel Meeting Agendas and Presenters

WASHINGTON, D.C.
FEBRUARY 16-17, 2011

Call to order and welcome
John Ahearne, Chair

Overview of the study task and origins and the National Academies' study process
Sarah Case, Study Director; John Ahearne, Chair

IFE committee briefing to the panel on expectations
Gerald Kulcinski, Inertial Fusion Energy Committee Co-Chair

Review of charge to the panel, the U.S. Department of Energy's interests in the committee and panel reports, and nuclear weapons proliferation risks for an inertial fusion energy program
David Crandall, Office of the Under Secretary for Science, U.S. Department of Energy

Indirect drive target physics at the National Ignition Facility (NIF)
John Lindl, Lawrence Livermore National Laboratory (LLNL)

Direct drive target physics at the Naval Research Laboratory (NRL)
Andrew Schmitt, NRL

Direct drive target physics at NIF
David Meyerhofer, Laboratory for Laser Energetics

Heavy ion target physics
John Perkins, LLNL

Z-pinch target physics
Mark Herrmann, Sandia National Laboratories (SNL)

Non-proliferation considerations associated with inertial fusion energy
Raymond Jeanloz, University of California, Berkeley

PLEASANTON, CALIFORNIA
APRIL 6-7, 2011

Welcome and call to order
John Ahearne, Chair

System considerations for IFE
Tom Anklam, LLNL

Overview of laser inertial fusion energy system and key considerations for IFE targets
Michael Dunne, LLNL

ALBUQUERQUE, NEW MEXICO
MAY 10-11, 2011

Welcome and call to order
John Ahearne, Chair

Inertial confinement fusion (ICF) targets at Los Alamos National Laboratory (LANL)
Juan Fernandez, LANL

Design and simulation of magnetized liner inertial fusion targets
Steve Slutz, SNL

ROCHESTER, NEW YORK
JULY 6-8, 2012

Welcome and call to order
John Ahearne, Chair

Welcome and overview of Laboratory for Laser Energetics (LLE) ICF program
Robert McCrory, LLE

Direct-drive progress on OMEGA
Craig Sangster, LLE

Polar drive target design
Radha Bahukutumbi, LLE

Facilitating NIF for polar drive
David Meyerhofer, LLE

Fast and shock ignition research
David Meyerhofer, LLE

LPI issues for direct drive
Dustin Froula and Jason Myatt, LLE

Heavy ion target design
B. Grant Logan, LBNL

Discussion of LIFE targets and program
Michael Dunne, LLNL

Technical feasibility of target manufacturing
Abbas Nikroo, General Atomics

WASHINGTON, D.C.
SEPTEMBER 20-22, 2012

Welcome and call to order
John Ahearne, Chair

Development of the technologies for laser fusion direct drive
John Sethian, NRL

Overview of current NRL program for ICF/IFE
Steve Obenschain and Andrew Schmitt, NRL, and Frank Hegeler, Commonwealth Technology at NRL

Overview of LPI physics and LANL understanding
David Montgomery, LANL

Understanding of LPI and its impact on indirect drive
Mordechai Rosen, LLNL

Assessment of understanding of LPI for direct drive (solid-state)
Dustin Froula, LLE

Assessment of understanding of LPI for direct drive (KrF)
Andrew Schmitt, NRL

State of the art for LPI simulation
Denise Hinckel, LLNL

C

Acronyms

AEC	Atomic Energy Commission
ASC	Advanced Simulation and Computing
CBET	cross-beam energy transfer
CH	carbon-hydrogen (plastic used as an ablator)
CTBT	Comprehensive Nuclear Test Ban Treaty
DCA	Detailed Configuration Analysis
DD	direct drive
DOE	U.S. Department of Energy
DPSSL	diode-pumped, solid-state laser
DT	deuterium-tritium
EP	extended performance
EU	European Union
FCT	flux-corrected transport
FES	Fusion Energy Sciences
FI	fast ignition
FIRE	Fast Ignition Realization Experiment
HAPL	High Average Power Laser (Program)
HDC	high-density carbon

HIF	heavy-ion fusion
HiPER	High Power Laser Energy Research (facility)
IAEA	International Atomic Energy Agency
ICF	inertial confinement fusion
ID	indirect drive
IFE	inertial fusion energy
ILE	Institute of Laser Engineering (Japan)
IR	infrared
ISI	induced spatial incoherence
LANL	Los Alamos National Laboratory
LBNL	Lawrence Berkeley National Laboratory
LEH	laser entrance hole
LIFE	Laser Inertial Fusion Energy
LLE	Laboratory for Laser Energetics
LLNL	Lawrence Livermore National Laboratory
LMJ	Laser Megajoule Facility (France)
LPI	laser-plasma interactions
LTE	local thermal equilibrium
MagLIF	magnetized liner inertial fusion
MAGO	Explosively Driven Magnetic Implosion (Russia)
MCF	magnetic confinement fusion
MTF	magnetized target fusion
NAS	National Academy of Sciences
NIC	National Ignition Campaign
NIF	National Ignition Facility
NNSA	National Nuclear Security Administration
NPT	Nonproliferation Treaty
NRC	National Research Council
NRL	Naval Research Laboratory
ORNL	Oak Ridge National Laboratory
PD	polar drive
PEP	project execution plan
RTL	recyclable transmission line

SI	shock ignition
SNL	Sandia National Laboratories
SNM	special nuclear materials
SRS	stimulated Raman scattering
SSD	smoothing by spectral dispersion
THD	tritium-hydrogen-deuterium
TPD	two-plasmon decay
VNIIEF	All-Russian Research Institute of Experimental Physics
YOC	yield over clean